P9-CSG-633

ECHINODERMS

Biological Sciences

Editor

PROFESSOR A. J. CAIN
MA, D.PHIL

Professor of Zoology at
the University of Liverpool

ECHINODERMS

David Nichols
MA, D.PHIL
Professor of Biological Sciences
at the University of Exeter

HUTCHINSON UNIVERSITY LIBRARY
LONDON

HUTCHINSON & CO (*Publishers*) LTD
178–202 Great Portland Street, London W1

London Melbourne Sydney
Auckland Bombay Toronto
Johannesburg New York

First published 1962
Second (revised) edition 1966
Third edition 1967
Fourth (revised) edition 1969

The paperback edition of this book is sold
subject to the condition that it shall not,
by way of trade or otherwise, be lent, re-sold,
hired out, or otherwise circulated without the
publisher's prior consent in any form of
binding or cover other than that in which it is
published and without a similar condition
including this condition being imposed on the
subsequent purchaser

© David Nichols 1967 and 1969

This book has been set in Times, printed in Great Britain
on Smooth Wove paper by Anchor Press, and
bound by Wm. Brendon, both of Tiptree, Essex

09 065993 7 (cased)
09 065994 5 (paperback)

'Non coelo tantum, sed et mari suae stellae sunt'

J. H. LINCK, *De Stellis Marinis*, 1733
(probably the first book about echinoderms)

CONTENTS

FIGURES

PREFACE TO THE FOURTH EDITION

The revision for this latest edition has been more extensive than previous ones. In particular, the surge of interest in fossil echinoderms engendered by the publication of those parts of the *Treatise on Invertebrate Paleontology*[12] dealing with the non-crinoid echinoderms has meant that the taxonomic framework of the book has had to be changed, the treatment of the fossil groups revised and additional theories on relationship included. I am particularly glad to be able to change my reconstruction of a cystoid pore-rhomb (Fig. 12) in the light of recent work[74]: my first effort, based on inadequate descriptions, has been convincingly shown to be sadly misconceived. On the other hand, the enormous increase in our knowledge of very early echinoderms has not caused me to change my ideas on the origin of the group.

On the neontological side, the work of my own group and associates, both past and present, has caused a change in my ideas on the mode of operation and evolution of the all-important water-vascular system, and I have taken the opportunity to make alterations in sections dealing with this.

1969   D.N.

PREFACE TO THE FIRST EDITION

While the present-day members of the phylum Echinodermata form an assemblage of animals remarkably distinct from all others, it is highly probable that in their early history they, or something very like them, provided a spring-board to chordate organisation. For this reason, and also because echinoderms are such common members of both the littoral fauna and fossil-bearing rocks, the phylum is an important one for students of zoology and palaeontology. It is inevitable that in presenting a comprehensive account of a very varied group of animals within the confines of a small book some aspects are treated in greater detail than others; I am fully aware, too, that generalisations are made all too often. But my aim has been to present an introductory survey of the basic structure, adaptive radiation and evolutionary history, as far as it is known, of the living and fossil groups; to treat at greater length some problems which have a general biological interest; and to view the phylum in perspective among the rest of the animal kingdom.

All who have studied the echinoderms will know the remarkable contribution to the field by Dr L. H. Hyman in her series *The Invertebrates*[8]. So comprehensive is her review of echinoderm anatomy, development and geographical distribution that I have treated these aspects in the briefest way only. The collective work *Physiology of Echinodermata*[11] is also a deep and thorough treatment and I include no more than the briefest summary of echinoderm physiology in consequence. Instead, I have concentrated on presenting a picture of the general nature of this outstanding phylum and in doing so I have dealt at some length with the fossil history, for I believe that nowhere is there a concise summary of the evidence and interpretations of the unique fossil record of this fascinating group, and I offer this as my excuse for dwelling on the subject for several whole chapters and parts of others.

It is a pleasure to express my thanks to those colleagues who have so kindly helped me by reading the manuscript and suggesting corrections and improvements. In particular I thank Professor A. J. Cain, Mr A. C. Campbell, Miss Ailsa Clark, Dr J. D. Currey, Dr D. Heddle, Mr R. C. Higgins, Dr K. A. Joysey and Dr J. D. Woodley.

I

INTRODUCTION AND GENERAL FEATURES

Among the echinoderms we find some of the most beautiful of all sea-creatures. Who has not marvelled at the symmetrical beauty of the starfish and brittle-star, the mozaic perfection of the spherical sea-urchin, or, for those fortunate enough to have seen it, the ballet of the swimming feather-star? While the star shape or the globular theca are the most common body forms, some members of the phylum look remarkably worm-like and others medusoid. What, then, are the features which separate the echinoderms as a phylum from superficially similar animals?

Everybody knows a starfish (Fig. 1b), and will probably have seen it creep, mouth downwards, rather slowly over the sea-bottom. At first sight it might appear that the five arms bend to produce movement, but a starfish can move with little or no lateral flexure of its arms. On the underside are five grooves, radiating from the mouth to the tips of the arms, and each groove contains soft appendages, the *tube-feet*. These are the organs which move the animal, by extending and bending in concert, and which provide one of the main places on the animal where exchange of respiratory gases takes place. Round the spherical sea-urchin, too, one can see five columns of tube-feet (Fig. 1e)—in fact, it is as though the tips of the starfish's arms had been held above the centre of the animal and the arms sewn together. Often, the sea-urchin's tube-feet can be seen extended free in the surrounding water and waving gently to improve respiratory efficiency. In the brittle-stars (Fig. 1d) the tube-feet are scarcely visible among the spines of the arm, though they are important for providing pur-

chase during locomotion and for feeding, and, in some, for
burrowing. In the feather-stars (Fig. 1a) and sea-lilies (Fig. 3g)
there are columns of tiny tube-feet on the upper side of the arms,
that is, on the same side as the mouth (*oral surface*), but in this
case they are not used for locomotion, since the animal, if it moves
at all, always keeps the oral surface uppermost. In the sea-
cucumber (Fig. 1c) five columns of tube-feet lie along the worm-
like body, and those on the underside may help to move the
animal. Generally speaking, the tube-feet are the most important
effector organs these animals possess, as will be shown in a later
chapter (p. 135).

Another feature which one notices immediately about most
echinoderms is that they possess a skeleton. The ability to build a
skeleton of a particularly strong kind (p. 123) in the mesodermal
layer of the body has undoubtedly played a large part in the
success of the group. The calcite skeleton is not referred to as a
shell, because it is internal and always has living tissue outside
it; so it is called a *test*.

From what has been said so far, it is clear that three features
of the echinoderms are outstanding: first, the adult members show
a body pattern having structures present in fives (*pentamery*);
secondly, the animals all possess *tube-feet*; and thirdly, they all
have a *calcite skeleton*, even though it may be reduced, as in
holothuroids. The first feature, while nearly universal in present-
day forms, has not always been so: the fossil history of the group
(a history which is known as well as, if not better than, that of
any other group of animals) shows that from their first appearance
there were several attempts at non-pentamerous forms, none of
which got very far. Today we see departures in the sea-cucumbers
and in some of the urchins; but even here, where the animal has
become secondarily bilateral, many of the basic structures are
laid down in fives, which is rather different from the early non-
pentamerous forms, in which there was no hint of basic pentamery
in the arrangement of parts of the body and where some of the
parts were in addition totally asymmetrical.

Surprisingly, it was not until the middle of the nineteenth
century that the echinoderms were finally seen as a phylum on
their own, distinct from the other principal radiate group, the
Coelenterata (the polyps and medusae). The anatomists pointed
out that whereas the coelenterates have a simple gut with a single
hole, the echinoderms have an alimentary canal with mouth and

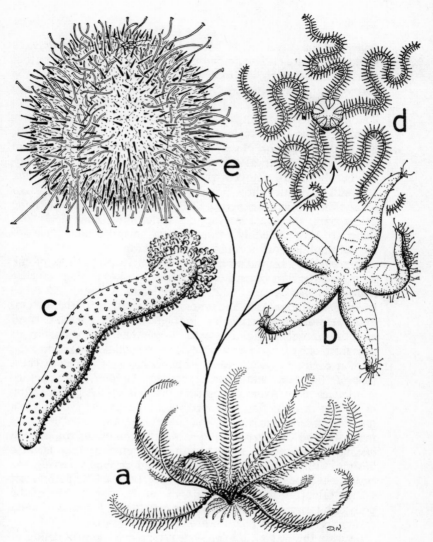

Fig. 1 Examples of the five living classes of echinoderms

a The feather-star *Antedon*, Class Crinoidea, closest to the ancestral form.
b The starfish *Asterias*, Class Asteroidea.
c The sea-cucumber *Holothuria*, Class Holothuroidea.
d The brittle-star *Ophiothrix*, Class Ophiuroidea.
e The sea-urchin *Echinus*, Class Echinoidea.

anus, and the organs of the body lie in a distinct body cavity: the echinoderms are *coelomate*. Soon after this it became evident that there are two fairly distinct ways in which a coelom can be formed: either it can arise as a split in the mesoderm (*schizocoely*) or as an outgrowth of the gut cavity or enteron (*enterocoely*). The echinoderms belong to the latter group, and this removes them from the schizocoelous annelids, molluscs and arthropods, and relates them more closely to that group of the animal kingdom generally referred to as the *minor coelomates*, that is, the sipunculids, phoronids, ectoprocts, brachiopods, hemichordates, etc. It is still not known how fundamental the two modes of coelom formation are, because there are a few echinoderms (notably some brittle-stars) and some brachiopods which show schizocoely, but these are now generally recognised as being peculiar developmental modifications for an unknown reason.

Superficially, the tube-feet look like mere projections of the animal's outer skin, but in fact they represent blind-ending branches from a highly complex and little-understood system of tubes which lies in the coelom. The tubes are actually *part* of the coelom, and the walls are permeable enough to allow coelomic fluid and the amoeboid floating cells it contains (*coelomocytes*) to pass from one cavity to another. We can recognise two parts of the tubular coelomic system, each probably having to some extent a vascular function (that is, transportation of food and excretory substances to and from the tissues of the body), though this is by no means certain: first, the *water vascular system*, whose main function is concerned with the activity of the tube-feet; and secondly, the *perihaemal system* which usually surrounds the haemal tissue, which is a system of interconnecting lacunae dealing with certain waste products and probably having the additional function of producing coelomocytes. The perihaemal system often lies to one side of some of the main nerves of the body, and some writers prefer to call it the *hyponeural sinus system* to reflect this relationship.

Because the water vascular system relies on the production of hydraulic pressure to operate the tube-feet, it is not surprising that in an animal with a body wall that is rendered rigid by a skeleton there would need to be provision for equalisation of internal and external water pressures; for instance, an animal like a sea-urchin, attached to a rock in a shallow sea, will be subjected to much greater ambient pressure when the tide is in than when it is out. This pressure equalisation probably occurs at the *madreporite*, a hole or series of holes connecting the outside sea-water with the water vascular system, and, in some, with other parts of the coelom too (see also p. 41). The madreporite can clearly be seen on the upper surface of most starfishes and sea-urchins, and is the one external structure which breaks the symmetry of these creatures.

The water vascular, perihaemal and haemal systems, and indeed the nervous system too, have the same basic format: a ring-shaped vessel round the mouth (the *circum-oral* component) and five *radial* vessels arising from it to supply the arms. In addition, there is usually an *axial* component arising from the circum-oral ring and leading towards the aboral side of the body.

Some of the early echinoderms of the Cambrian period of geological time built themselves a cup-shaped body or *theca* with projecting food-catching processes, each having a ciliated groove on one side, and probably also a skeletal supporting rod. In this arrangement of the body both mouth and anus are directed upwards and such a body form has persisted to the present day in the sea-lilies and feather-stars (the CRINOIDEA), but in all the other present-day forms the mouth no longer points upwards: in the starfishes (ASTEROIDEA), brittle-stars (OPHIUROIDEA) and sea-urchins (ECHINOIDEA) the mouth is directed downwards, while in the sea-cucumbers (HOLOTHUROIDEA) it is directed forwards. Such changes in attitude have inevitably meant changes in feeding habit: whereas the crinoids feed exclusively on particles of organic matter suspended in or falling through the waters around them, the others nearly all collect food from the sea-bottom. There are almost as many different ways in which they do this as there are species, but in general the primitive asteroids and most of the ophiuroids sweep material along the underside of the arms into the mouth; the holothuroids, too, mostly sweep the area in front of them with special tube-feet. But some advanced asteroids, with special suckers on their tube-feet for greater purchase, have evolved a method of preying upon molluscs to augment their diet of tiny particles. Then in the primitive echinoids we see the occurrence of teeth, used, again in conjunction with strongly suckered tube-feet, to rasp organisms from the rocks over which they move, while more advanced echinoids burrow through the substratum and take in particles of it by means of special tube-feet, then digest off the adhering organic matter as the particles pass through the gut.

No special excretory organs are found in this phylum, and the process of ridding the body of metabolic waste is something of a mystery. There is little doubt that the coelomocytes play a major role, since they have the power to move through the soft tissues, ingesting material as they go, and to discharge themselves and their contents to the exterior. The axial organ plays a part too (p. 67).

In all but a few freak forms, the sexes are separate. There is good evidence that most primitive echinoderms had a single gonad opening by a single pore close to the anus. In present-day crinoids, however, the gonads are not contained in the main part of the body, but occur in large numbers in side-branches of the arms. In the asteroids, ophiuroids and echinoids there are five gonads within the main part of the body, each opening by its own pore, while in the holothuroids there is a single gonad only. Fertilisation takes place in the sea. Maturation of the germ cells is carefully timed so that a spawning animal of either sex can induce neighbouring animals to follow suit, greatly increasing the chances of fertilisation. In addition, it seems that many echinoderms are gregarious, to their obvious advantage in reproduction. Development in the great majority is indirect; that is, they pass through a larval stage. Since the adults are somewhat sluggish, the larvae are the main dispersive phase of the animal and remain in the plankton for sufficient time to be swept from the place of their birth to new areas, or to restock the original areas. In addition to their dispersive function, the larvae will aid the species in feeding from a different source from their adults, and thus when food is short larvae and adults will not compete.

Apart from being a phylum with a world-wide distribution and with considerable diversity of structure on the basic plan, the echinoderms occupy a unique place in zoology: it is generally agreed nowadays that in them, or at least very close to them, we have the ancestors of the great phylum of the Chordata, at the head of which Man himself sits. The way in which these two superficially quite dissimilar phyla are related is one of the most exciting speculations in the whole field of zoology, and the manner of their connexion is dealt with at the end of the book.

It now remains to consider the groups which make up the phylum Echinodermata. The most useful first sub-division is into four *subphyla* (see pp. 172–175): the CRINOZOA, mostly attached to the sea-bed when adult; the ASTEROZOA, the star-shaped forms; the ECHINOZOA, free-living globular or worm-like; and the HOMALOZOA, wholly extinct, non-radiate forms.

The present-day echinoderms are placed in five easily-recognised *classes* within three of the subphyla:

CRINOZOA

1. CRINOIDEA, the sea-lilies and feather-stars, the most primitive living class and the only crinozoan class with living members. Greek 'lily-like'.

ASTEROZOA

2. ASTEROIDEA, the starfishes (or sea-stars). Greek: 'star-like'.
3. OPHIUROIDEA, the brittle-stars and basket-stars. Greek: 'snake-like'. These and the asteroids are sometimes combined as the Class Stelleroidea, when they themselves become sub-classes.

ECHINOZOA

4. ECHINOIDEA, the sea-urchins, sand-dollars and heart-urchins. Greek: 'spiny'.
5. HOLOTHUROIDEA, the sea-cucumbers. Greek: 'violent expulsion', referring to their defensive evisceration.

There are wholly extinct classes in all the sub-phyla except the Asterozoa:

CRINOZOA

6. CYSTOIDEA, sessile cyst-like forms, regarded by some as the simplest (but not necessarily the most primitive) echinoderms. Greek: 'bladder-like'.
7. EOCRINOIDEA, superficially similar to the cystoids, but possessing a different sort of respiratory organ. Greek: 'early crinoid'.
8. PARACRINOIDEA, with some crinoid-like features and some cystoid-like. Greek: 'near the crinoids'.
9. PARABLASTOIDEA, with some cystoid and some blastoid features. Greek: 'near the blastoids'.
10. BLASTOIDEA, sessile bud-like forms. Greek: 'bud-like'.

ECHINOZOA

11. EDRIOASTEROIDEA, mostly free forms, though some later ones acquired stems; certain features are crinozoan-like. Greek: 'sessile starfish'.
12. HELICOPLACOIDEA, cigar-shaped forms with spiral plating. Greek: 'spiral-plated'.
13. OPHIOCISTIOIDEA, free forms with huge, plated tube-feet on the underside. Greek: 'snake capsule'.
14. CYCLOCYSTOIDEA, disk-like forms with an outer ring of large plates that may have borne large tube-feet. Greek: 'round capsule'.

HOMALOZOA

15. HOMOSTELEA, spoon-shaped forms with a stem but no feeding arm. Greek: 'same stem', referring to the similarity of the homalozoan stem to that of other crinozoans.
16. HOMOIOSTELEA, asymmetrical flattened forms with feeding arm and stem. Greek: 'similar stem', referring to an attempt to distinguish a slight difference between the stem of this group and that of the homosteles.
17. STYLOPHORA, asymmetrical flattened forms with a feeding arm but no stem. Greek: 'pillar-bearing'.

Finally, there are three groups of undecided category containing little-known fossils which are undoubtedly echinoderms, but which are not yet seen to fit easily into the presently-recognised subphyla:

18. 'HAPLOZOANS', radially and biradially symmetrical forms consisting of a solid, dome-shaped theca. Greek: 'simple animals'.
19. CAMPTOSTROMATOIDEA, medusiform echinoderms not unlike pelagic holothuroids. Greek: 'flexible-framed'.
20. LEPIDOCYSTOIDEA, bud-shaped forms somewhat similar to camptostromatoids and found in the same beds. Greek: 'scaly cystoid'.

It is fairly certain that the original echinoderm was rather different from those which move freely, though somewhat slowly, over the bottoms of modern seas. It probably sat on or in the sea-bottom while feeding on the rain of detritus falling to it from the waters above. It may have had a cup-shaped protective test, and food-collecting arms or brachioles supported by a skeleton. The living class whose members most closely resemble this archetype is the Crinoidea, so we shall consider this class first in the survey of the living forms which follows. Then, three chapters are devoted to the totally-extinct groups, and four to the more important general features of the phylum. Finally, a speculative account is given of the phylogeny of echinoderms and their relationships in the animal kingdom.

2

THE CRINOIDEA

Conveniently enough, the crinoids (sea-lilies and feather-stars), in addition to being very similar to some of the first echinoderms to appear in the fossil record, have retained a primitive structure throughout their evolutionary history, and there are living representatives which are quite easily obtained for study. So they make the obvious starting-point in our brief survey of the classes. The fossil record tells us that the stemmed crinoids flourished in the Palaeozoic, particularly in the Carboniferous period, some rocks of that age consisting of little more than the fossil remains of crinoids. Since then the stemmed forms declined in importance and were superseded by advanced crinoids, the comatulids or feather-stars, that become secondarily free during their life-history; these are today far more widespread than their less mobile forbears, particularly in coastal waters. The best-known example, *Antedon*, is a comatulid, though its basic anatomy is very similar to that of stalked forms.

General body plan

The typical crinoid body is made up of a *stem* with some sort of holdfast, a cup-like calyx or *theca*, housing the lower part of the body, a domed flexible roof, the *tegmen*, housing the rest of the body, with the mouth at its centre, and a series of *arms* arising from the sides of the theca; each arm usually bears a large number of *pinnules*, branching alternately from its sides.

The stem, seldom exceeding two feet in modern forms but sometimes reaching seventy feet in the past, consists of a vertical

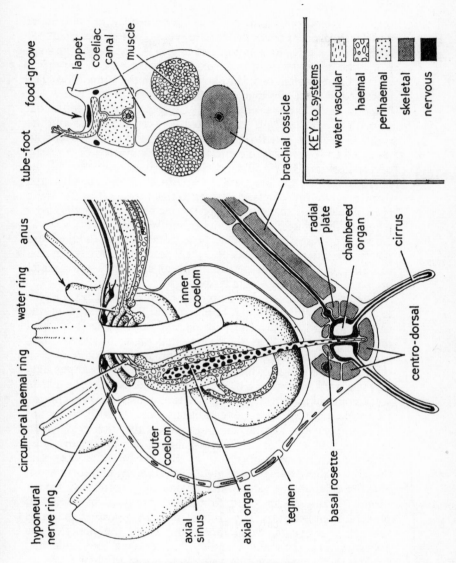

Fig. 2 Basic anatomy of a crinoid

Diagrammatic vertical section through the disk and part of one arm of a crinoid, based on the comatulid *Antedon*, and, on the right, transverse section through one arm.

series of round or star-shaped ossicles embedded in epithelium, and bears branches or *cirri* which are retained on the very reduced stem of comatulids for temporary fixation. At the top of the stem is a single plate, the *centro-dorsal*, forming the base of the theca. This is the only stem ossicle to be retained in the comatulids. The sides of the theca are made up of two whorls of five plates, the *basals* aborally and the *radials* orally. In some forms an additional whorl, the *infrabasals*, is intercalated between the basals and the centro-dorsal. In *Antedon*[20] the thecal plates are very reduced and modified to form a capsule enclosing the chambered organ (part of the coelom, see p. 30) and the most important part of the nervous system; the five basals are reduced, fused together and form a *rosette* which acts as a cover to this capsule (Fig. 2). There is fossil evidence that primitively the tegmen contained a single whorl of five plates, and in some primitive living forms, such as *Hyocrinus*, these are still visible close to the mouth, though extra plates are intercalated between them and the radials. In less primitive forms the five plates disappear and small extra plates are scattered over the whole tegmen.

The arms arise from the boundary between the calyx and the tegmen—actually, they are borne on the radial plates of the calyx. There are five arms primitively, and this number is retained in a few forms, such as the living *Ptilocrinus*; but more often each arm divides into two, and sometimes this forking is continued again and again, the number of arms reaching about sixty in some stalked forms, e.g. *Metacrinus*, and even up to 200 in some comatulids, e.g. *Comanthina*. The number of arms appears to be determined by food conditions, and Clark[21], who has monographed the existing crinoids, makes the interesting point that the number of arms in comatulids seems to be correlated with the depth and temperature of the seas in which they live, those in shallow warm water tending to have around forty arms, and those in deep, colder water having only ten; those having twenty to thirty arms tend to occupy intermediate depths of moderate temperature. The skeletal pieces supporting the arms are called *brachials*, while those supporting the pinnules are *pinnulars*. The arms and pinnules represent the food-collecting system, outclassing anything seen elsewhere in the living echinoderms; each pinnule has a groove on its oral side leading into a similar groove in the arm which bears it, and the arm groove in turn leads into a system of grooves in the surface of the tegmen which converge on the mouth. All these grooves are bordered by an alternating series of cover-plates, the *lappets,* and groups of tube-feet arise from the sides of the grooves just inside the lappets. The arrangement and activity of these tube-feet is discussed fully in Chapter 12; here, suffice it to say that their function appears to be to collect food from the rain of detritus falling upon the animal, bundle it up in strings of mucus and convey it to the food grooves. When the animal is disturbed by some creature that might nibble the delicate tube-feet, they retract into the groove and are immediately covered by the lappets. Presumably, while they are retracted the food already in the groove system can continue its journey to the mouth.

As is general in almost all sessile echinoderms, and indeed other invertebrates which feed on the rain of detritus falling from the waters above, the mouth is situated as near the centre of the food-collecting area as possible and the anus is displaced to the side, slightly upsetting the radial symmetry. Discharge of faecal waste, always a problem in ciliary feeders, is dealt with by having the anus mounted on a flexible spire, so that the waste can be aimed away from the grooves; some advanced comatulids even go so far as to send the food grooves on a circuitous route to the mouth, to avoid passing too close to the anus; others, as though admitting defeat, even dispense with food grooves on those arms which arise adjacent to the anus.

In all living echinoderms the mouth is always surrounded by five (or multiples of five) special tube-feet usually having a mainly sensory function. In crinoids, as in most of the others, these tube-feet arise from a circum-oral water vascular ring vessel. Their turgor is maintained by the pressure in the water vascular system, and this, in conjunction with longitudinal muscles, enables them to bend to taste food passing towards the mouth. The mouth leads into a short oesophagus and this into an intestine which either has a single loop with many diverticula, or makes four turns and lacks diverticula. There is a short rectum, whose walls are muscular and which can pulsate, probably causing water to be taken into the rectum, either as an enema or for anal respiration, a process which certainly occurs in some other echinoderms (p. 76).

Reproductive organs
We saw in Chapter 1 that in early echinoderms there was almost invariably a single gonopore opening on the side of the theca,

inferring that there was a single gonad only, and this was housed in the theca. But in the crinoids the reproductive system is entirely different: there are many gonads, situated on the proximal pinnules of each arm (except the first pair, which usually act to protect the tegmen) and extending to well over half the pinnules of the arm. Only by close examination are the sexes distinguishable. The gametes are shed into the sea by rupture of the pinnule wall and usually development proceeds in the water, or sometimes while the egg is attached to the parent by a sticky secretion, unless a brood pouch is present on the arms (e.g. *Notocrinus*) or on the pinnules (e.g. *Isometra*).

Excretory system

It is most likely that the main method of excretion, as in other echinoderms, is by coelomocytes. In crinoids the full coelomocytes collect in special spherical *sacculi*, which lie in rows alongside the ambulacral grooves, and it has been said that the sacculi discharge their waste into the sea at intervals. An interesting thing about these organs is that they also seem to collect pigment granules from the body tissues when the animal dies, so that in preserved specimens they appear darkly coloured.

The coelom

In all echinoderms the coelom is subdivided very early in development (p. 150) into parts destined to become, on the one hand, the perivisceral coelom (that surrounding the main organs of the body) and, on the other, the so-called 'tubular coelomic systems', the water vascular and perihaemal (or hyponeural) systems. The perivisceral coelom continues into the arms, in each of which it becomes the *coeliac canal*. This canal continues also into the pinnules. In the walls of all the coelomic canals are tiny *ciliated pits*, which may be the places where coelomocytes pass into the surrounding tissue, or where they aggregate during the excretory process.

The tubular elements of the system are as follows:

(1) *The water vascular system.* Although of similar plan to that of other echinoderms (p. 18), this system in crinoids has no direct connexion with the exterior, that is, there is no madreporite as such. Instead, there is a large number (150 in *Antedon*) of tiny ciliated water pores in the interradii of the circumoesophageal water ring connecting the water vascular system with the perivisceral coelom; in addition, the tegmen is pierced

by many *ciliated funnels*, so that in effect there is a connexion between the water vascular system and the outside sea-water *via* the coelom, though whether water actually passes between the two has not been established. The oral water vascular ring gives off a pair of oral tube-feet in each interradius, which probably test the food before it enters the gut, and a radial water-vessel in each radius, dividing where necessary to pass into each arm; then branches go to each pinnule, and the canals in arms and pinnules give off the branches to the tube-feet. It seems to be general among crinoids that the water vascular canals, including the circum-oral ring, are traversed by muscle fibres, the function of which is to produce the necessary pressures for tube-foot extension, as dealt with in Chapter 12.

(2) *The perihaemal system.* As its name suggests, this system normally surrounds the haemal complex, though some authors prefer to call it the *hyponeural sinus system*, to emphasise its relation to one part of the nervous system. The perihaemal system, like the water vascular, has a circum-oral ring, an axial part called the *axial sinus,* and a radial branch to each arm which becomes divided into two coelomic arm canals. There is doubt, however, as to whether this system is homologous with the perihaemal system of other echinoderms, or represents merely another part of the perivisceral coelom. As it is said to be confluent with the axial sinus, it is here regarded as a part of the perihaemal system, though the 'axial sinus' itself may not be homologous with the cavity so named in other groups.

In addition to these two tubular coelomic components, there is, in common with other echinoderms, a system of interconnecting lacunae in a spongy tissue called the *haemal system.* This system sends elements to all parts of the body, and consists of a number of rings surrounding the first part of the gut, an axial portion, and a branch to each arm which sends 'twigs' to each pinnule and tiny channels to each tube-foot. Round the first part of the oesophagus is the *circum-oral haemal ring* which is in communication by many channels with the *genital haemal ring* just below it, from which the *radial haemal strands* arise; these, passing along each arm just above the coeliac canal, give off branches to the pinnules, and it is these branches, greatly expanded, which contain the gonads (hence the name of the ring round the gut from which they arise). Also arising from the genital haemal ring is the so-called *spongy body*, an extensive system the main part of which is in close association with the gut walls. The spongy body is thickest near the 'hub' of the animal, that is, at the axial complex, where it surrounds an *axial organ*. We can conveniently consider this organ as part of the haemal system throughout the phylum, because it is always in close association with it, the channels of the haemal system ramifying within it. Aborally, the axial organ continues further than the spongy body (Fig. 2), disappearing, in fact, only after it has passed through the rosette formed by the five fused basal plates. During development the axial organ sends a process along the middle of each radial haemal strand to the pinnules, processes which later enlarge to form the gonads. The connexion between organ and processes is broken later in life, but many early workers thought that the germ cells originated in the organ, and hence called it the genital stolon. Other workers reported seeing it pulsate (as do most of the tubular coelomic parts, anyway) and because of this called it the heart.

In addition to these three tubular coelomic systems, present in all echinoderms, there is in crinoids another coelomic structure aborally, called the *chambered organ*. This occupies the region of the body enclosed by the thecal plates (Fig. 2) and sends branches to the cirri.

So far, no function has been mentioned for the perihaemal system, the chambered organ or the haemal system (the first two coelomic, the third mesodermal). Indeed, this is one of the outstanding problems of echinoderm physiology. We can but remark that the perihaemal canals and the chambered organ both lie in close association with some of the main nerves (e.g., in crinoids the radial nerve cords have the radial perihaemal vessels and the aboral nervous system has the chambered organ), and it may be that their activities are functionally related. In the case of the haemal system, certain parts of it have been shown to exercise an excretory function, as will be mentioned in connexion with another class later (p. 67). But very little is known about those parts of the system that are intimately related to the gut and gonads, with a connexion between the two. It may be that the system has a special part to play in nourishing the developing gametes: it may provide a 'nutritional clear-way' between gut and gonads.

The nervous system

One of the features of the echinoderm nervous system is that a mainly sensory nerve plexus lies just beneath the external epithelium of almost the entire body. In crinoids, this is not as extensive as in other classes, though it is prominent locally where special sensitivity is required, as, for instance, at the tips of some tube-feet and on the cirri. In addition to this general plexus there is an extensive 'central' system also. In crinoids this is in three main parts, each inter-communicating. The main sensory part lies orally, and has been called the *superficial oral* or *ectoneural* system; it is little more than an expansion of the basi-epithelial plexus in a ring round the mouth and beneath the centres of the ambulacral grooves. Just below (aboral to) this is another sensory system, the *deeper oral* or *hyponeural* system, also consisting of a ring round the gut and branches in each arm; this time there are two to each arm and pinnule, and they lie laterally. Then the main motor system in crinoids, concerned with posture and movement[22], is the *aboral* system, which surrounds the chambered organ within the thecal plates and has a ring embedded in the radial plates; the ring gives off a brachial nerve to each arm and pinnule, situated in a canal within the brachial or pinnular ossicles. Lateral con-nexions run to the lateral brachial nerves of the deeper oral system.

We may summarise the basic anatomy of crinoids as follows: they have a stem by which they are attached to the sea-floor for at least part of their life-history; their food is collected by an extensive system of arms and pinnules and passed by cilia to an upwardly directed mouth. Circulation of essential materials and excretory products is probably carried out by the coelomic systems which ramify through the animal.

Evolution and adaptive radiation of the crinoids

Despite the multiplicity of Palaeozoic forms, we still have a very incomplete picture of early crinoid evolution[23], and no picture at all of crinoid origins; there are no convenient 'intermediate forms' between the various crinozoan classes to clarify the picture. The oldest true crinoid is *Ramseyocrinus* (Fig. 3*a*) from the Lower Arenig Beds of the Lower Ordovician of South Wales[25]. The material of this form is rather imperfect, but we can see that there are whorls of basal and radial plates, that the arms consist of single columns of brachials which branch several times, and that the stem is a four-lobed structure as wide as the theca. But it does not provide an ancestor for later crinoids, because it has only four arms, the fifth apparently being replaced by a short papilla carrying the rectum and anus. Presumably the ancestor of the more regular Palaeozoic crinoids antedated this form.

Ramseyocrinus probably belongs to the Order INADUNATA, because of its rigid thecal plates. When the crinoids appear in quantity, a little later in the Ordovician, two other orders can be recognised; the FLEXIBILIA, with flexible theca, and the CAMERATA, with rigid theca but no radianal plate and with some of the proximal arm ossicles incorporated into the theca.

Though these three Palaeozoic orders are generally agreed to have been distinct, certain evolutionary trends are common to all of them. For instance, the primitive crinoid theca is conical in shape, with straight sides, as in *Ramseyocrinus*: during evolution the base of the cup becomes flat, then concave, then finally surrounds the top of the stem. The ratio of height to width decreases.

Some authorities[1] consider the number of whorls of thecal plates below the ring of radials, either one (monocyclic) or two (dicyclic), to be important taxonomically, but most other workers consider that this arrangement is misleading, because this character appears again and again in the group, and in some genera, e.g. *Uintacrinus*, some species are monocyclic while others are dicyclic[7], so that Bather's[1] sub-classes Monocyclica and Dicyclica can profitably be rejected.

Among early inadunate crinoids are forms such as *Hybocystis*, Ordovician (Fig. 3*d*), in which only three of the five ambulacra

Fig. 3   Evolution and adaptive radiation in the crinoids

a and *c*, Inadunata.
a Ramseyocrinus (L. Ord.), the earliest crinoid to appear in the fossil record.
b Petalocrinus (Sil.), with arms expanded into food-collecting flanges. Arm nearest reader removed.
c Hybocystis (Ord.), in which two of the ambulacra pass down the theca instead of being borne on arms.

d and *e*, Camerata.
d Platycrinites (M. Sil.—M. Perm.), with irregularly branching arms.
e Barrandeocrinus (M. Sil.), in which the pinnules bend to form food-collecting troughs. The two arms nearest the reader removed.

f and *g*, Recent Articulata.
f Ptilocrinus, a stalked form.
g Antedon, a comatulid.

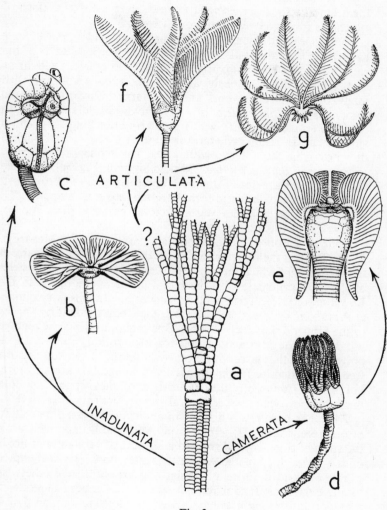

Fig. 3

are continued on to arms: the food grooves of the other two ambulacra continue down the sides of the theca. In other forms, e.g. *Petalocrinus*, Silurian (Fig. 3*b*), the arms are expanded to form fan-like food-gathering structures, over the oral surface of which the food grooves ramify extensively.

There is less radiation among the flexible crinoids[23], most forms having the typical lily-like appearance. To mention just two, the earliest flexibles are *Protaxocrinus* (Middle Ordovician) with arms that branch dichotomously and *Ichthyocrinus* (Silurian) with many plates in each whorl and each arm bifurcating. Of the camerates, a typical form is *Platycrinites*, Middle Silurian to Middle Permian (Fig. 3*d*), with irregularly branching arms and a spiral stem. In many camerates it appears that the arms sagged down over the theca, like a wilting lily. This tendency is taken to an extreme in *Barrandeocrinus*, Middle Silurian (Fig. 3*e*), in which the arms hug the sides of the theca and the pinnules of each arm bend towards each other to form a trough into which, apparently, water and food particles were admitted between the pinnules, then taken to the mouth.

None of the three foregoing orders passes into the Mesozoic. In the Triassic the fourth and last order, ARTICULATA, appears, in the members of which the tegmen is flexible and the basal ossicles of the arms articulate with the radial plates of the theca. In the Lower and Middle Triassic all articulates have stems (an example is *Pentacrinites*), but stemless forms, the Comatulida, first appear in the Upper Triassic of Mexico. This group, in which advantage is taken of the arm articulation for locomotion, appears to be polyphyletic; that is, several groups within the Articulata became free independently. In the Jurassic the comatulids are still rare, but in the Cretaceous they flourish, and they appear to hold their superiority, at least in littoral waters, until the present day.

There has been much discussion about the systematic position of one recent form, *Hyocrinus* from Antarctic waters, which has the straight-sided conical theca reminiscent of very early crinoids of the order Inadunata. At least one textbook[7] links it, via a couple of Jurassic forms of similar appearance, to the otherwise totally Palaeozoic Inadunata, but recently it has been suggested[23] that its primitive features have arisen by convergence, and that it really belongs, like all other living crinoids, to the order Articulata.

3

THE ASTEROIDEA

The starfishes, the first of the two living asterozoan classes we shall consider, are famous objects of the seashore and infamous visitors to oyster-beds, where their ability to wrap themselves round the bivalves, tug at the shell against the muscles which close it and extrude their stomachs to digest away the meat has exasperated fishermen and fascinated zoologists for years. In the early days of oyster-farming the fishermen would drag a dredge across the beds to collect the starfish, tear them in pieces and throw them back again; they did not know then that there was a very good chance of every piece regenerating lost parts to become a complete starfish again, considerably increasing the menace in time.

During the early part of their life (p. 152) the asteroids become fixed to the sea-bottom for a short time, and this has been considered good evidence for regarding them as fairly close to the Crinozoa; during evolution they have managed to break free and invert. There are also features in their adult anatomy which indicate their rather primitive position (p. 162).

Though by far the majority of starfishes are pentamerous, the class exhibits greater departure from this number of rays than any other. For instance, the Atlantic species *Luidia sarsi* has the normal five arms, while *L. ciliaris* has seven; the Sunstar, *Solaster*, may have any number from fifteen to fifty, and the number increases with age.

General body plan

Basically, the asteroid body consists of a central disk with the mouth in the middle of the undersurface and anus in the centre of the upper. A number of arms project laterally. The delimitation of arms and disk is not very marked; indeed, in some, such as the Duck's Foot, *Anseropoda* (Fig. 5*j*), the outline is roughly that of a pentagon. The main part of the gut, the stomach, is situated in the disk (Fig. 4), but two canals with side pouches extend into each arm. Similarly, the central parts of the tubular coelomic systems are found in the disk, with radial extensions along each arm. All asteroids have a skeleton of internal plates, but while in some they may be closely packed and form a complete layer in the body, in others some of the ossicles, particularly those of the upper surface, have large spaces between them. Round the mouth is a special *peristomial ring* of closely abutting plates, two *ambulacrals* in each radius alternating with two *adambulacrals* in each interradius. Oral to each pair of adambulacrals there is usually a single plate, called the *interradial*. The elements of this ring represent the first of the series which extend out along the oral surface of the arms, each arm having two parallel columns of closely packed ambulacrals with a single column of adambulacrals on each side of them. This explains why the adambulacrals cannot be called interambulacrals—each pair round the mouth is split between two adjacent radii. The ambulacral plates normally form a deep groove in the oral surface of each arm and the plates have pores between them, often in a zig-zag, taking canals from the tube-feet to ampullae associated with them (see Chapter 12); they seldom, if ever, bear spines. The adambulacrals, adjacent to them, bear strong spines which in life can touch the substratum or bend inwards to protect the ambulacral groove. Lateral to the adambulacrals come the plates forming the sides of the arm: the *inferior* and *superior marginals* and, where present, the *dorso-laterals*. Then finally come the *carinals*, forming the mid-line of the aboral surface, though these too are absent in some. In one primitive group of starfishes, the Phanerozonia, the marginals are conspicuous and closely set (it is this fact to which the subclass name, meaning 'visible edge', refers), while in the more advanced forms (formerly united as a group, Cryptozonia, 'hidden edge') the marginals are no bigger than the other lateral and aboral plates.

Little blisters of the body wall, *papulae*, formed of the ciliated external and coelomic epithelia, with a thin connective tissue and muscle layer sandwiched between them, bulge through the spaces

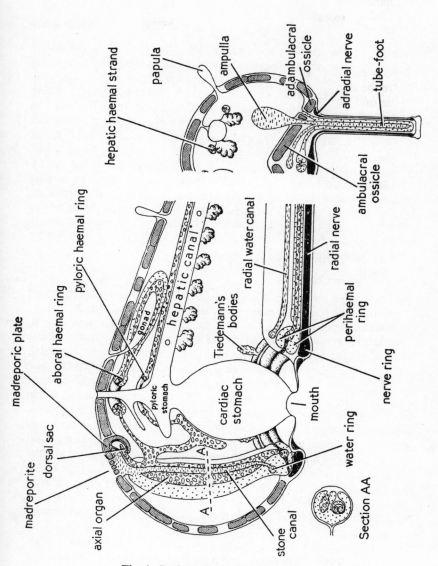

Fig. 4 Basic anatomy of an asteroid

Diagrammatic vertical section through the disk and part of one arm
of a starfish, based on *Asterias*. On the right, transverse section through
one arm, showing one tube-foot and its ampulla. A section of the axial
complex is shown at lower left.

See Fig. 2 for key to systems.

between the ossicles, so that, in addition to the oxygen which gets into the water vascular system through the tube-foot walls, gaseous exchange takes place directly between the sea-water and the perivisceral coelomic fluid via the papular wall; the papulae have a further function to do with excretion (p. 43). In some phanerozones the papulae are restricted to isolated regions, the *papularia*, on the aboral surfaces, usually close to the proximal end of each arm, as in *Pectinaster*.

The whole system of ossicles is flexible, and the shape of the body can be altered by special skeletal muscles. Most important are the upper, lower and lateral transverse ambulacral muscles on the oral side of the arms, which are for altering the depth of the grooves, and the circular and longitudinal muscles over the whole body just below the coelomic epithelium, which are for altering the shape of the body.

The alimentary canal

The mouth, surrounded by an uncalcified peristome stretched across the space within the peristomial ring of ossicles, leads into a vertical gut (Fig. 4). The first part of this, a short oesophagus, may have ten pouches in its walls reminiscent of those of the crinoid gut (not shown in Fig. 4). The next part, the stomach, has a wide *cardiac* portion, held by radial mesenteries to the wall of the disk, followed above by a smaller pyloric portion, from which ten *hepatic canals* arise, two to each arm. Each canal has a large number of heavily-sacculated *caeca* arising from its sides, and it is here that digestion is continued and absorption takes place. There is a two-way current along the canals to carry food to the caeca and digestive juices to the stomach. Apart from the expected mucus and enzyme secreting cells, the walls of the caeca contain glycogen storing cells, the contents of which disappear in starved animals[26]. Lastly, there is a short *rectum*, usually with a branching diverticulum, the *rectal sac*. The gut may end blindly in some of the burrowers.

The stomach is very flabby and manœuvrable. When feeding on an animal that is too large to be taken in, the starfish will extrude its stomach by turning it inside-out; it can then wrap it round the prey and secrete enzymes on to it to digest the meat. Even when a small shell is taken into the body, the gut wall will become closely applied to it, so that the secreted enzymes are used economically.

The coelom

The general body cavity is not subdivided as it is in so many other echinoderms; the perivisceral coelom of the disk is a single cavity, continuous with that in the arms. The water vascular, haemal and perihaemal systems are well developed and have probably been studied in more detail in this group than in any other.

The water vascular system

As usual, the main job of this system is to supply the fluid necessary for the hydraulically operated tube-feet and to maintain it at the required pressure. One of the plates on the aboral surface of the disk, derived from the ring of interradial plates formed first of all (p. 158), is pierced by a large number of very small pores, together called the *madreporite*; part of this plate curves towards the interior of the disk to form a cranny in its inside surface, the importance of which we shall see when we consider some of the other tubular coelomic systems. The madreporite leads down into a nearly vertical tube bearing the somewhat unjustified name *stone canal*, because early workers found its walls strengthened with tiny spicules against which their instruments tinkled and grated as they dissected. The fossilised remains of these ossicles have even been found in the earliest asteroids, the Somasteroids[44]. The canal is not a simple tube but usually contains a scroll-shaped projection from one side of its inner wall (Fig. 4,*AA*) to increase the surface area of the ciliated epithelium lining it, and thus the strength of the current in it. At its oral end the stone canal opens into a *circum-oral water ring*; at its aboral end it is said[37] to open to the surrounding axial sinus and also to have a tiny projection, the *madreporic ampulla*, within the cranny of the madreporic plate. Arising from the wall of the circum-oral vessel are five pairs of interradial *Tiedemann's bodies*, thought to secrete some of the coelomocytes. In some asteroids (though not in the common *Asterias*) the vessel also bears *polian vesicles* interradially, which are muscular sacs with the probable function of maintaining turgor in the system.

From the water ring (internal to the skeleton) arise the radial water canals (external), one to each arm; the canals pass through the skeleton just outside the peristomial ring, and the pores through which they pass can be seen in some early fossils. Each canal passes along its arm above (aboral to) the bridges formed by the lower transverse ambulacral muscles; between these muscles, on alternate sides, branches are given off to the tube-feet, and each tube-foot has an ampulla internal to the skeletal pieces. A muscular valve at the point where the branch from the radial canal joins the lumen of the tube-foot isolates each tube-foot/ampulla unit from the rest of the water vascular system so that the necessary hydrostatic pressure can be built up. The structure and activity of the tube-feet and their ampullae is dealt with more fully in Chapter 12. Each radial canal ends at the arm-tip as a terminal tentacle, basically a tube-foot with no ampulla, whose function is mainly sensory.

One other structure, not really part of the water vascular system, can conveniently be dealt with here—the *dorsal sac* (Fig. 4). This tiny isolated vesicle shares the cranny in the madreporic plate with the madreporic ampulla. Projecting into it is part of the haemal system (see below).

The arrangement of the structure in the cranny of the madreporic plate suggests that the whole system here may act as a pressure sensitive organ. The tissue between the madreporic ampulla and the dorsal sac is stretched fairly tightly across the cranny, and any change in pressure within the so-called ampulla which might be caused by, for instance, tidal rises and fall, may deform the stretched tissue at the expense of the dorsal sac, and nerves in the periphery of the stretched tissue may well detect the deformation and initiate appropriate action by the tube-foot/ampulla systems and perhaps the haemal system too. That there is some method of counteracting the effects of a change in the 'head' of water above the animal is clear from experiments by Fechter[104], but so far no experimental evidence of a pressure-regulating function for the madreporic complex has been obtained.

Haemal system

As usual, this has a number of ring elements round the gut connected together by an axial part, and radial components leading off into each arm. The *oral haemal ring*, running in a septum in the perihaemal ring (p. 41), gives off the *radial haemal strands*, also in a septum, which pass up each arm external (oral) to the radial water vessels. Round the pyloric stomach is the *pyloric haemal ring*, giving off branches, the *gastric haemal tufts*, to the walls of the cardiac stomach, and four branches, the *hepatic haemal strands*, to the walls of the hepatic caeca of each arm; these gastric parts of the haemal system are not enclosed in perihaemal elements. The *aboral haemal ring*, round the rectum, gives off two branches to each arm, leading to the gonads. In the asteroids it is apparently not possible to distinguish a separate axial haemal system (equivalent to the spongy body of crinoids) from the *axial organ*—the two seem to be one structure. This lies to one side of the stone canal and sends an aboral process, the *head-piece*, into the dorsal sac within the cranny of the madreporic plate. The function of the head-piece is unknown, though it might be connected with the detection of external pressure on behalf of the haemal system and axial organ.

Perihaemal system

This system surrounds the oral, aboral, radial and axial parts of the haemal system. The *oral perihaemal ring* round the oesophagus sends a *radial perihaemal canal* into each arm; both ring and canals are subdivided by a septum (Fig. 4) so that the ring consists of inner and outer portions, and the canals of two elements side by side. The oral haemal ring and the haemal strands lie within this septum. The radial perihaemal canals, like the haemal strands, are said to give off branches to the tube-feet. From the outer haemal ring arises the *axial sinus*, surrounding axial organ and stone canal. The sinus and the stone-canal communicate aborally. The *aboral perihaemal ring* round the rectum and, of course, surrounding the aboral haemal ring, is not in communication with the axial sinus but is a separate cavity giving off branches to the gonads.

As mentioned before (p. 18) some writers prefer to call this system the *hyponeural sinus system*, to emphasise its position with respect to parts of the nervous system, to one side of which most parts of it lie.

The nervous system

It is important to remember that in all echinoderms a thin nerve plexus, the *basi-epithelial plexus*, lies within and below the entire epithelium covering the body; in places it is thickened into fairly definite tracts, though these are anything but separate nerves by vertebrate standards.

The nervous system of the asteroids has received far more attention than that of other echinoderm groups, chiefly through the work of Smith[18, 41, 42, 43, 116] and Kerkut[106, 107].

Originally, as in crinoids, three separate systems were recognised: ectoneural, hyponeural and entoneural, but more recently it has become evident that the hypo- and entoneural systems are continuous, so that a division between these two is inappropriate. It would seem better to divide the whole system into *superficial* and *deep* parts. The superficial part is mainly sensory; it is composed of the general basi-epithelial plexus and those parts of it which are specially thickened to provide through-conducting paths. The chief of these are, first, the *circumoral nerve ring*; secondly, the *radial* (=*perradial*) *nerve cords* in the mid-line of the underside of each arm; and, lastly, the *adradial nerve cords*, two to each arm, which pass along the underside of each arm lateral to the tube-foot/ampulla canal (Fig. 4). Any special regions of the body, such as special ciliary tracts or aggregations of sensory receptors, normally have a thickening of the plexus just beneath them.

The *deep* component is predominantly motor. It consists of segmentally arranged centres, internal to the superficial plexus, and various nerve tracts to effector organs. The main centres lie adjacent to the tube-foot/ampulla junctions: first, there are *Lange's nerves* (better called *Lange's centres*) just aboral to the lateral parts of the radial nerve cord. These are aggregations of twenty or more neurones lying in the floor of the radial perihaemal canals, separated from the nerve cords below by a thin layer of connective tissue across which nerve fibres pass to join the two systems. Nerves pass from Lange's centres to the inferior transverse ambulacral muscles just above them, and to the tube-foot/ampulla system. Secondly, there are *motor centres* lateral to the tube-feet, innervating the ampullae (which thus have a double innervation) and in addition sending lateral motor nerves up the inside of the body wall just below the coelomic epithelium. These nerves supply various skeletal muscles. In those forms with two lobes to each tube-foot ampulla, such as the phanerozone *Astropecten*, the lateral motor centres innervate the lateral lobes while Lange's centres innervate the medial lobes.

Besides the general scattering of neurosensory cells over the asteroid body, there are five light-sensitive *optic cushions*, one at the base of each terminal tentacle. Each cushion contains numerous ocelli in the form of cups, the external part of which is sometimes modified into a lens. The walls of the cup consist of cells containing a red pigment and, interspersed between them, retinal cells with nerve fibres passing into the nearby radial nerve cord.

The ambulacrum
From what has been said about the radial parts of the tubular

coelomic, haemal and nervous systems, it can be seen that they all lie *outside* the skeletal pieces of the ambulacrum, though of course covered by an external epithelium. This is termed an *open ambulacrum*, and is the type first seen in the Crinozoa and many early echinozoans. In the other three living classes, however, the skeletal pieces close over the radial elements of the various systems as further protection, thus forming a *closed ambulacrum*. The nature of the ambulacrum has been used to support phyletic relationship[8] but it will become evident later (p. 162) that the echinoderm ambulacrum has closed over independently at least twice.

Some aspects of asteroid physiology

Respiration. It has already been said that oxygen from the sea-water is taken into the body via tube-feet and papulae. In some phanerozone starfishes special aggregations of papulae are present in grooves between some or all of the marginal plates of the arms. These are called *cribriform organs*. Cilia drive water in an oral direction through these organs, and it appears that the strong currents are also used to bring food into the ambulacral grooves from the dorsal (aboral) surface. In other phanerozones a respiratory chamber may be present on the aboral surface: special spines called paxillae (more fully described in Chapter 11) can close over to lie parallel to the aboral body wall and abut against their neighbours to form a chamber between them and the body. Papulae lie in the floor of this chamber, and water, drawn in and out by a combination of muscular contractions of the body wall every ten to twenty seconds and ciliary action, passes over them for gaseous exchange. The chamber is also used by some starfishes for brooding the larvae.

Excretion. Like other echinoderms, the asteroids lack a definite excretory system; but injection of coloured particles has revealed that these, and presumably other waste products too, are ingested by coelomocytes, some of which pass into the papulae and are extruded by rupture of the papular wall, while others apparently pass through the external epithelium of certain parts of the body, particularly the disks of the tube-feet, to be extruded as slime.

Osmotic regulation. There seems to be none: the body fluids are almost isotonic with sea-water. The ionic tolerance varies astonishingly, some starfishes being intolerant of more than the most minor salt level change, while others enjoy much greater freedom, and hence ecological range (e.g. *Asterias rubens* which ranges from the North Sea at $30^{0}/_{00}$ salinity to the Baltic at $8^{0}/_{00})^{27,\ 28}$.

Feeding. The primitive asteroids, as one would expect from their forbears, were probably ciliary feeders, relying on currents over the animal's surface to collect food particles, passing them in strings of mucus to the ambulacral grooves and thence to the mouth; other particles would probably be picked up from the surface over which the animal moved, possibly by the tube-feet, and added to the same stream. Then the tube-feet acquired suckers, probably for better locomotion in the first place; and later the ability to prey on bivalves was seized upon. Considerable interest has been focused on the problem of how starfishes open bivalves. It has been shown recently[39] that the combined pull of the starfish tube-feet and body muscles on a bivalve's adductor muscle may be as high as 3,000 gm. It is commonly said that the starfish wins the tug-of-war because it can keep up a persistent pull of this magnitude for hours on end. This does not seem to be the case: between five and ten minutes is all the time required for the starfish to pull the shells open about 0·1 mm, which is sufficient to insinuate the stomach and start digestion; and the force is relaxed and reapplied at intervals until the adductor muscle has been rendered ineffective.

Recently, it has become clear that echinoderms as a whole may be able to acquire food by two quite distinct agencies: first, there is the usual method of digesting other organisms in the gut and transporting the useful substances round the body in the coelom; secondly, they can absorb free amino acids from the surrounding sea-water directly into the outer tissues of the body, a method which must assume great importance in nourishing such external effectors as spines, tube-feet and pedicellariae, some of which have no connexion whatever with the coelom and hence with a supply of gut-derived foodstuffs. The whole epithelial surface, including that of the effectors mentioned above, is revealed by electron microscopy to possess a velvet-like 'pile' of microtubules which are projections from the epithelial cells. Experiments with labelled amino acids and glucose in the medium in which starfishes are kept have shown[35] that the uptake is an active process, since these substances can be extracted from very dilute solutions; it is fairly rapid on first immersion but gradually decreases over a period of eight hours or so. On present evidence, it seems that the process is mainly to nourish those parts of the body that are outside the encasing skeleton: very little (less than 1%) finds its way into internal organs, even after some weeks. Free amino acids do occur in sea-water at concentrations of up to 125μg/l, and it is possible that more is released into the vicinity of the starfish by the digestive activity of its own stomach, especially when extruded (p. 38).

The regulation of gamete release. When a starfish in an aquarium tank spawns, others in the tank will follow after a short while. Both sexes will respond to the shedding of either sex. In animals with external fertilisation, the advantage in this 'follow my leader' activity is obvious. Experiments have indicated[120] that gamete release is regulated by neurohormonal activity of substances apparently secreted from the radial nerve. Extracts prepared from the radial nerves of starfishes have been shown to bring about, first, contraction of the gonad wall, secondly, the completion of maturation of the gametes and, thirdly, the release of the eggs from the extracellular substance which holds them in a clump when in the ovary. The process appears to be more than a simple triggering effect, however, since there is a half-hour lag between the injection of shedding-substance and gamete release. It seems likely that in addition to the shedding substance there exists in the nerves an inhibitor to it, called 'shedhibin', the concentration of which fluctuates during the annual reproductive cycle and becomes maximal just before the time of spawning. Spawning is a function of the balance between shedding substance and shedhibin, though what causes the decrease in the activity of the inhibitor at the time of spawning is not known. Also unknown is the pathway by which the shedding substance reaches the gonads.

The evolution and adaptive radiation of the Asteroidea

The views expressed here are contrary to those[8] giving the ophiuroids a closer phyletic connexion with the echinoids than to the asteroids. There would seem to be excellent grounds, particularly from palaeontological evidence, for regarding the group Stelleroidea (asteroids and ophiuroids) as natural, and it is logical to regard it as a class within the Echinodermata at the same taxonomic level as the Crinoidea, Echinoidea and Holothuroidea. However, so deeply engrained in common usage are the terms Asteroidea and Ophiuroidea, and so comparatively distinct are the living members, that it would seem inappropriate to relegate them to the rank of sub-class, which is the suggestion of some workers[32, 44]. The earliest starfishes show characters consistent with their being considered ancestral to both the later starfishes and the ophiuroids, and these early forms are often included in a separate sub-class, the Somasteroidea. From their overall body plan and the nature of the ambulacra, the somasteroids more closely resemble the asteroids than the ophiuroids, and therefore I treat them here as a sub-class of the Asteroidea, but it should be made clear that some palaeontologists consider the three groups of star-shaped echinoderms as of equal status.

The earliest members of the Somasteroidea are *Chinianaster* (named after the locality) and *Villebrunaster* (after its collector, Villebrun) from the Upper Tremadoc and Lower Arenig beds of the Lower Ordovician of St Chinian, in the south of France[32, 44]. The specimens were found embedded in flint-like nodules lying in the soil of vineyards. The owners of the vineyards generally cleared the nodules from the soil for convenience of cultivation and left them in spoil-heaps at the edge, a rich source for the fossil collector of all sorts of animal remains. A third somasteroid is *Archegonaster*, from the Upper Arenig beds of the Lower Ordovician (slightly later than the other two) of Bohemia. Specimens of this form were also preserved in nodules. An interesting feature of preservation, which also seems to throw light on the behaviour of the animals when they lived, is the fact that generally only the disks and proximal parts of the arms are found in the nodule, and very often the arms are flexed away from the oral surface. Spencer[44] suggests that the animals were burrowers, and that they lived in a manner similar to many recent asteroids and ophiuroids, that is, with only the arm tips protruding above the surface of the substratum; then, only those ossicles which were actually buried at death were held together sufficiently long for fossilisation to occur. In this case preservation occurred by the precipitation of minerals round the animal. Subsequently the calcite of the skeleton was dissolved away, so that most of the studies have been made on casts; nevertheless, the impressions left behind contain fine detail of spines, joints and muscle facets, in fact, almost as much as could be hoped for from a modern asteroid skeleton. So our knowledge of this basal group is surprisingly full.

The most primitive somasteroid, as well as one of the earliest, appears to be *Chinianaster* (Fig. 5*a*, *b*). This form has a large disk with a wide pentagonal peristome and five petaloid arms. Along the mid-line of each arm on the oral surface are two columns of alternating ambulacral ossicles, each shaped like a half-cylinder, which do not contain a groove between them. Instead, a cylindrical channel, presumably the site of the radial water vascular canal, runs between the two columns, totally enclosed. In this feature they are unlike the true asteroids, but, as we shall see, the channel tends to become open on the oral face in later forms. Branches from the enclosed canal passed out on alternate sides to deep concavities in the oral faces of each ambulacral ossicle. It appears highly probable that each concavity was the site of insertion of a tube-foot with a small head-bulb contained in the concavity and acting like an ampulla. It should be mentioned that this is the only known early asteroid without pores between the ossicles for the passage of the canal to an ampulla. The ambulacral columns diverge at their inner ends to become the edge of the peristome; the proximal member of each column, which comes to abut against its neighbour from the next arm at the adjacent inter-radius, is slightly larger than the rest of the series, paralleling the enlarged *ad*ambulacrals which occupy the same position in the mouth region of later, true, starfishes. The rest of the skeleton on the underside consists of many longitudinal series of elongated ossicles, the *virgalia*, arranged at an angle to the ambulacrum of each arm. The skeleton of the aboral side consists of small triradiate spicules in the form of a meshwork.

In the somewhat less primitive but contemporaneous somasteroid *Villebrunaster* the arms are still petaloid in shape. The main anatomical advances towards the true asteroid condition are, first, that the outermost virgalia of each column are shortened and expanded laterally to form a series of *marginals*, outlining the plan-view of the animal, and, secondly, that there are pores lateral to the ambulacrals, that is, between them and the first virgalium of each series, which most probably carried a canal from the tube-foot to an *internal* ampulla (Fig. 5c). In some somasteroids a madreporite has been found lying in one of the oral interradii, but it is not clear if that was its original position or whether it had fallen from the aboral site.

The latest and morphologically most advanced of the fossil somasteroids, *Archegonaster*, Upper Arenig (Fig. 5d), shows the following further advances: first, the ambulacral channel is open more widely on the oral surface than in *Chinianaster*; secondly, there is a series of adambulacral ossicles, possibly modified virgalia, lying lateral to the ambulacrals and articulating with them in an ophiuroid-like manner; and, thirdly, the virgalia themselves are restricted to the distal ends of the arms. This last modification leaves the interradial parts of the disk apparently free of structural ossicles, though there may have been loose ones embedded in a membrane. The marginals are very much stouter than they are in *Chinianaster*, and it is clear in this form that the madreporite is aboral.

Until the end of 1961 the somasteroids were thought to have become extinct in the Lower Ordovician, but then some material originally described in the 1870's of a recent starfish called *Platasterias latiradiata* from the seas off West Mexico was examined and found to have somasteroid characters[32, 33]. A full report of this 'living fossil' is still awaited, but the preliminary examination has revealed that the somasteroid virgalia appear to be homologous with the crinoid pinnular ossicles, and this, of course, suggests a direct connexion between the crinoids and the asteroids.

We will now briefly follow the evolution of the true asteroids from the later somasteroids. The virgalia are lost or transformed completely, and we see now the attainment of a second row of marginals on the aboral side of the edge, the *supramarginals* (the original marginals now being called the *inframarginals*). We also see that the ambulacral ossicles expand laterally from being

Fig. 5 Evolution and adaptive radiation in the asteroids

  a The first asteroid, a somasteroid *Chinianaster* (L. Ord.); oral view of skeleton.
  b to h, diagrammatic transverse sections of the arms of a fossil series of asteroids showing the probable evolution of the ossicles.
  b *Chinianaster* (L. Ord.).
  c *Villebrunaster* (L. Ord.). The outermost virgalia have become the marginals, and the tube-foot ampullae are internal.
  d *Archegonaster* (L. Ord.). The innermost virgalia have become the adambulacrals.
  e *Platasterias* (Recent), a living somasteroid, in which the virgalia are somewhat telescoped.
  f *Petraster*, the first true asteroid (L. Ord.).
  g *Hudsonaster* (Ord.).
  h *Xenaster* (Dev.).
  i to l, typical British asteroids, showing the main trends.
  i *Astropecten*, a phanerozone which burrows in sand or gravel.
  j *Anseropoda*, a spinulosan with thin body, also a burrower.
  k *Asterias*, a forcipulate which lives on the surface.
  l *Solaster*, a spinulosan.

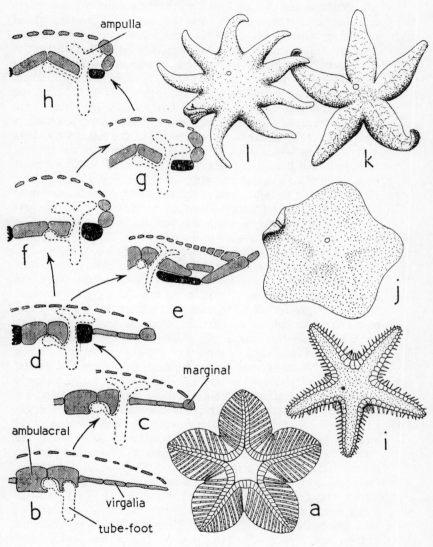

ampulla

h

g

f

e

d

marginal

c

ambulacral

b

virgalia

tube-foot

a

l

k

j

i

Fig. 5

barely more than half-cylinders in the somasteroids to the true asteroid condition of elongation at right angles to the long axis of the ambulacrum. Lastly, we see the migration of the pores carrying the ampullar canals from their ancestral position lateral to the ambulacral plates, to the true starfish position, between adjacent ambulacral plates. In the post-somasteroid starfishes the madreporite may be aboral, marginal or oral in position in closely related forms, so this feature does not have any phyletic significance.

The first true asteroid is probably *Petraster* (Fig. 5*f*) from the Lower Arenig beds of the Lower Ordovician. This is roughly contemporaneous with the somasteroid *Chinianaster*, but there is no doubt that *Petraster* is a true member of the PHANEROZONIDA. Among the various species of this genus one can trace the gradual elaboration of the supra-marginals from a rather vague series of ossicles close to but not touching the inframarginals along part of the arm to a definite series articulating with the inframarginals all along it. The ambulacral plates are rather poorly known, but there already seems a tendency for lateral expansion. The ambulacral groove is not yet well marked.

In the next stage, typified by *Hudsonaster*, Ordovician (Fig. 5*g*), the main advances are the enlargement of the two rows of marginals into really strong elements for support of the body, the flexure of the ambulacral ossicles to form the typical asteroid ambulacral groove, and the strong attachment of the adambulacrals to the ambulacrals, as in later asteroids. Another interesting point is that in *Hudsonaster* the plates of the apical disk (in the centre of the aboral surface) show the typical echinoderm condition of a central plate surrounded by a ring of five interradial *genital* plates, one of which is the madreporite. Such an arrangement is seldom found in recent asteroids; in those forms which do have a pentameric arrangement of apical plates, such as *Tosia*, there is generally another ring of radial plates inserted between the central and the genitals. However, many starfishes (e.g. *Asterina*) are known to pass through the hudsonasterid condition when the aboral plates are first formed. Finally, in *Hudsonaster* one can see the adoption of the typical later asteroid condition in the mouth-frame: the proximal adambulacral plates of each column are larger than the rest and form the typical mouth-angle plates in each interradius round the mouth, thus paralleling the condition in the somasteroids, where the 'mouth-angle plates' were formed by enlarged ambulacrals.

The order PLATYASTERIDA, though it appears slightly later than the Phanerozonida (Middle rather than Lower Ordovician), has characters, such as the lateral expansion of the adambulacral plates, which suggest that it may be a more primitive order. Interestingly, the recent genus *Luidia* from British waters would appear to belong to this order, and may therefore, like *Plastasterias*, represent another 'living fossil' starfish[32].

But to return to the Phanerozonida, in *Hudsonaster* we have reached a condition of the skeleton more or less common to all subsequent asteroids. It now remains for the ambulacral plates to become markedly broader than long, and for the pores of the tube-foot ampullae to move nearer the mid-line of each ambulacrum. The first signs of this latter advance are seen in the Devonian *Xenaster* (Fig. 5h), in which the pores enter the interior of the arm at the lateral margins of the ambulacral plates, that is, where the ambulacrals and adambulacrals abut. In subsequent palaeozoic forms the tendency is for the pores to shift to a position about halfway along the ambulacral plates and at the same time for the ambulacral plates to oppose rather than alternate. This, the typical phanerozone condition, has persisted to the present day. Many of the phanerozones of Recent seas are burrowers. The most typical British form is *Astropecten* (Fig. 5i) of the sub-order Paxillosa, which normally burrows in sand or fine gravel, maintaining a connexion between the burrow and the surface of the substratum at the arm tips and sometimes at the disk as well. The papulae are restricted to the aboral surface, where they are protected and kept clear of falling sediment by the paxillae, the universal presence of which gives the sub-order its name. There are no suckers on the tube-feet. Probably the oddest phanerozones are the sphaerasterids, such as the deep water form from the South China Sea, *Podosphaeraster*, in which the body is almost perfectly spherical, and the ambulacra reach up to the equator, like the fingers of a hand holding an orb.

The British starfish *Porania* appears to lie close to the transition between the Phanerozonida and the Spinulosida. It is a non-burrower, and the tube-feet have sucker-disks for moving over the substratum, so it is probably best placed in the Spinulosida; in addition, there is a thick aboral integument for protection instead of the paxillate condition normally associated with the phanerozones.

The order SPINULOSIDA is not sharply separable from the

Phanerozonida. The order gets its name from the fact that aboral spines are normally present, usually arranged in groups either borne on a stalk ('pseudopaxillae'), as in the British form *Solaster* (Solasteridae) (Fig. 5*l*), or sessile on the surface, as in *Henricia* (Echinasteridae). The tube-feet of most spinulose starfishes are suckered. There are two British members of the spinulose family Asterinidae, *Asterina* and *Anseropoda* (= *Palmipes*) (Fig. 4*j*, and they show contrasting modes of life. *Asterina*, the small Gibbous Starlet, is a common shore animal, usually found in rocky areas. It is negatively geotactic, and presumably climbs to get closer to the more highly oxygenated surface waters. It is probable that the pull of the body on the tube-feet is the means of perception, because the effect can be reversed by tying a string round the body and pulling upwards. In *Anseropoda* the body is wafter-thin and the outline almost pentagonal, and normal movement is in the opposite direction— it is a burrower. It is peculiar as a burrowing form in that it possesses suckered tube-feet, and not the more usual pointed mucus-plasterers typical of burrowing phanerozones. Burrowing is apparently effected by musculo-skeletal methods rather than by digging with spines and tube-feet; the animal thrusts one side of its body beneath the surface of the substratum by pushing with the other. The animal is characteristic of shell-gravel, so the suckered tube-feet would be well suited to move the particles.

The order EUCLASTERIDA contains the very peculiar deep-water forms such as *Brisinga*, originally considered primitive and possibly close to the ophiuroids because of their possession of a distinct disk and long spiny arms, and, even more ophiuroid-like, very small ampullae to the tube-feet, and these partially embedded in the ossicles of the oral side. Perrier regarded the Brisingidae as the most primitive asteroids, because of their ophiuroid-like nature, but Fell considers that this group and the ophiuroids are convergent.

The last order, FORCIPULATA, is characterised by the presence of special pedicellariae (p. 130) with a basal piece and two valves (Fig. 22*g*, *h*). Here belong the asteriids such as *Asterias* and *Marthasterias*, the common starfishes of British seas (Fig. 5*k*), and the American Sunstar, *Heliaster* (not to be confused with the British spinulose form *Solaster*). These are the starfishes that have mostly adopted a predatory mode of feeding, by straddling molluscs and either pulling the shells apart or everting the stomach to digest away the meat.

4

THE OPHIUROIDEA

The members of this class, with much less variety of external form than any other, are delimited by their small disks and long spiny arms, though there is one exception (p. 52). It is these arms to which the class name refers ('snake-like'); the ease by which they break when handled has led to their common name 'brittle-stars'. Photographs of the sea-bottom often show huge aggregations of these animals, sometimes overlapping and even several deep, and this has given rise to the suggestion that they are the most successful of the echinoderm classes, judged on the criterion of numbers. Unlike most of the asteroids, the ophiuroids do not use only their tube-feet for locomotion, but usually move by sinuous flexures of the arms, one arm and the disk being thrust forward by oar-like movements of the two other pairs.

General body plan

The whole of the alimentary canal, the axial parts of the tubular coelomic and haemal systems and the gonads are contained in the disk (Fig. 6). The arms consist of little more than ossicles, their operating muscles and the radial ambulacral components. There is no anus, the star-shaped mouth on the oral surface serving for the removal of undigested food-remains. Primitively, the arrangement of plates on the aboral surface of the disk is a *central* plate, surrounded by a ring of five *radials* (Fig. 20c). More cycles of plates, usually multiples of five, then surround this central pattern to cover the disk. In more advanced ophiuroids this rather regular pattern of ossicles is obscured by reduction in

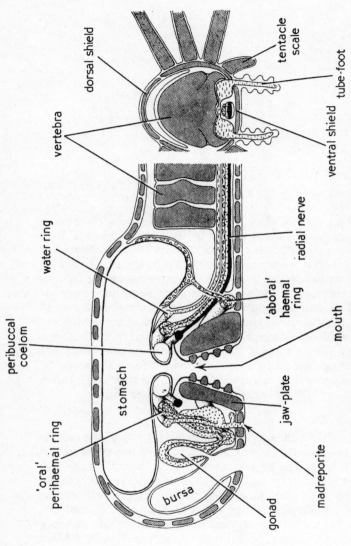

dorsal shield

tentacle
scale

vertebra

ventral shield   tube-foot

water ring

'aboral'
haemal
ring

radial nerve

peribuccal
coelom

'oral'
perihaemal ring

stomach

mouth

jaw-plate

bursa

gonad

madreporite

Fig. 6   Basic anatomy of an ophiuroid

Diagrammatic vertical section through the disk and part of one arm of a brittle-star, passing
through one genital bursa, one gonad, the axial complex and one ambulacrum. On the right, transverse section of an arm, passing through a pair of tube-feet.

See Fig. 2 for key to systems.

size of the original plates and the addition of other small plates. A pair of conspicuous plates, the *radial shields*, used extensively in taxonomy, lie close to the origin of each arm. On the oral side five interradial *oral shields* lie in the angles of the arms. These plates originate in the embryo as the second whorl on the aboral surface, that is, outside the ring of radials, and migrate to the oral surface; one of them is the madreporite, so they are most likely homologous with the basals of asteroids and the genitals (= basals) of echinoids. At the base of each arm there are a pair of *jaw plates* and a pair of *aboral shields*, homologous with the ambulacral and adambulacral plates respectively of the peristomial ring in asteroids.

The arms continue *inside* the disk on the oral surface. At their bases on each side is a slit, leading to a *genital bursa*, a sac which bulges up into the disk, the function of which is both respiratory and reproductive (p. 156). Each slit is bordered by a *bursal plate*. The main plates of the arms are a longitudinal series of ossicles called *vertebrae*, formed by fusion of two plates homologous with the asteroid ambulacral plates. Each hinges against its neighbour so that the arm can bend more in the horizontal plane than in the vertical; movement is brought about by two pairs of muscles, the *oral* and *aboral intervertebrals*. Surrounding the vertebrae are four series of plates—the *orals* in the mid-ventral line, the *aborals* mid-dorsally, and the *laterals* which bear spines. The tube-feet emerge from between the orals and laterals, and special modified spines on the laterals, the *tentacle scales*, lie at their bases.

Unlike the asteroids and crinoids, the ophiuroids have an ambulacrum the radial coelomic tubes and nerve cord of which lie internal to the skeleton (Fig. 6). This is called a *closed ambulacrum*, and in this feature they resemble the echinoids and holothuroids (p. 79), though its derivation from the open system of the early asteroids can be followed in the fossil record (p. 57.

The alimentary canal

Typically, the mouth opens into a sac attached to the aboral wall of the disk by strands representing broken-down mesenteries running across the coelom. This sac, usually called the stomach, has typically ten pouches bulging between the inside walls of the genital bursae. Only in one species so far known, the sub-tropical *Ophiocanops fugiens*, which is regarded by some[32, 33] as the most primitive living ophiuroid, are there any continuations into the arms.

The coelom

The body cavity is not nearly so extensive as in most other groups, because the genital bursae occupy those parts of the disk not filled by the gut. As in the crinoids and asteroids, the peri-visceral coelom of the disk is continuous with the coelom in the arms; here, however, so much space is taken up by the skeleton that the arm coelom is reduced to a narrow canal between the aboral and vertebral plates (Fig. 6).

The tubular coelomic and haemal systems

It is unnecessary to describe these in detail because they follow fairly closely the plan for the asteroids, with the difference that the madreporite of the water vascular system has moved second-arily to the oral surface, so that the axial components of the various systems run from their respective circum-oral rings, just above the jaw apparatus, *in an oral direction* to their 'aboral' rings; in consequence, the terms *oral* and *aboral* can no longer be used to denote position but only homology with the vessels in other classes. The stone canal gives off a madreporic ampulla just beneath the madreporic plate; the oral water ring, in addition to the five radial water canals, gives off four or five polian vesicles and ten branches which subdivide to become the lumina of the twenty buccal tube-feet, only ten of which are usually visible externally, the other ten arising close to the mouth. The ordinary tube-feet of the arms lie side by side, not alternately. In many phrynophiurids (p. 61) the axial organ of the haemal system consists of a lighter coloured 'oral' part and a darker 'aboral', the latter possibly homologous with the asteroid head-piece. The axial sinus of the perihaemal system is divided by the axial organ and stone canal into left and right parts, and possibly the left corresponds to the asteroid axial sinus and the right to the dorsal sac. Again as in asteroids, the sinus communicates 'aborally' with the stone canal.

Nervous system

The main nerve ring lies just oral to the axial sinus and gives off branches to each of the buccal tube-feet, which are sensory, and to each arm. These branches pass along the oral side of the vertebral ossicles either in a groove or in a canal. The basi-epithelial plexus, such an important feature of the nervous system of other echinoderm groups, is present only in the walls of the tube-feet, so almost all responses are of a generalised kind, ·involving transmission of excitation through the central pathways.

The origin of the ophiuroids

The problem of the origin of the ophiuroids is an outstanding example of how easy it is to draw conflicting pictures of phyletic relationship using evidence from different disciplines. It is probably true to say that if no fossil evidence were available a connexion between the ophiuroids and the echinoids would be most attractive, the similarity of body plan between ophiuroids and asteroids being purely convergent. Let us look briefly at the features in which the ophiuroids resemble the echinoids. First, there is the nature of the ambulacrum, a 'closed' system in both (p. 43), with the consequent presence of epineural canals in each radius and round the mouth; secondly, the tube-feet of both groups pierce the ambulacral plates rather than pass between them in the asteroid manner; thirdly, both have pluteus larvae (p. 153); and, lastly, there is some evidence of biochemical affinity: echinoids and ophiuroids are said to share the same type of sterol (cholesterol) while asteroids have a different type (stellasterol)[130]. It must be admitted that this is a pretty convincing set of similarities. Yet there are grounds for mistrusting them, even disregarding the fossil picture. For instance, it is shown (p. 156) that larval convergence has taken place in unrelated groups in answer to particular larval conditions, so that little weight can properly be placed on this feature. Again, it is becoming increasingly clear that the distribution of sterols is not nearly so clear-cut as originally thought[137], so the biochemical evidence, too, may be unreliable as a guide to phylogeny.

But, luckily, considerable fossil evidence is available, and this shows that without much doubt there is true phyletic relationship between asteroids and ophiuroids. The line from the somasteroid *Chinianaster* through *Pradesura*, *Palaeura* and *Stenaster* in the Ordovician to *Lapworthura* in the Silurian shows about as much of the establishment of those characters typical of present-day ophiurids as one could hope to get from the decidedly fragmentary fossil record of the early Palaeozoic. The series shows the gradual increase in size of the asteroid-like ambulacral plates until they become the vertebral ossicles of the ophiuroids; it shows the gradual envelopment of the radial water vessel until it becomes embedded in the substance of the vertebrae, and the gradual enlargement and migration of the adambulacrals until they become the lateral arm shields; round the mouth the transition from the asteroid mouth angle plates to the ophiuroid jaws can be followed, and from the asteroid first adambulacral to the ophiuroid adoral (lateral buccal) shields; and the madreporite in some of the early asteroids is oral, as it is in ophiuroids. Surely such a wealth of evidence can hardly fail to be convincing. Let us, then, look at the evidence of ophiuroid origin in more detail.

Judged on the criteria of general body form (the separation of disk and arms), the oral position of the madreporite and the ossicle arrangement in the arms, the first recognisable ophiuroid to appear is *Pradesura*[44] from the Lower Arenig beds of the Lower Ordovician (Fig. 7b). The oral surface of each arm has three types of plate: first, a double row of typical somasteroid-like alternating *ambulacrals* bordering a central channel for the radial water vessel and containing the typical hollows for tube-feet and their incipient ampullae shared between two adjacent plates: secondly, a column of longitudinally elongated *sub-laterals* lateral to each ambulacral column; and, thirdly, a similar column of rather

Fig. 7   Evolution and adaptive radiation in the ophiuroids

a Diagrammatic transverse section of the arm of *Chinianaster* (L. Ord.), the most primitive known somasteroid asterozoan, which is probably similar to the common ancestor of asteroids and ophiuroids. See also Fig. 5. The arrows represent the possible course of evolution.

b *Pradesura* (L. Ord.). The laterals and sublaterals have evolved from the somasteroid virgalia. Ambulacrals alternate.

c *Stenaster* (Ord.). Ambulacrals moving inwards and opposing; single enlarged lateral on each side.

d *Lapworthura* (Sil.). Arm becoming rounder and laterals taller.

e *Ophiocanops* (Recent). Primitive features include gonads serially repeated in dorsal cavity of arm, and gut canal in arm.
   i, T.S. of arm; ii, dorsal view of three 'segments'.

f *Asteroschema* (Recent), a euryalan. Ventral plate present.
   i, T.S. of arm; ii, the whole animal, showing unbranched arms and arm movement in all planes.

g *Ophiothrix* (Recent), an ophiurid. Ventral and dorsal plates present.
   i, T.S. of arm; ii, whole animal, showing arm movement mainly in horizontal plane.

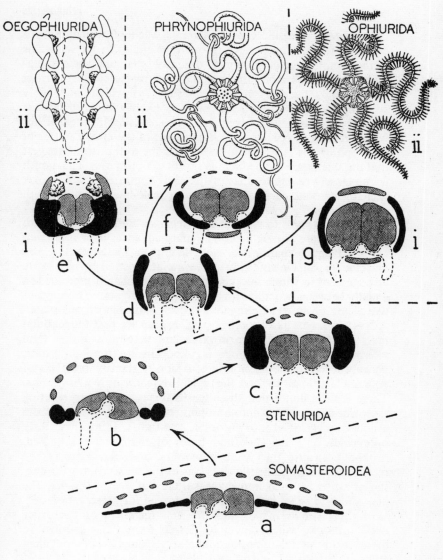

Fig. 7

larger *laterals* outside the sub-laterals, each of which bears a column of spines on a ridge set at right angles to the long axis of the arm, that is, in the typical ophiuroid manner. The ambulacrals closely resemble the homologous plates in the earliest somasteroid starfish *Chinianaster* (Fig. 7*a*) which is roughly contemporaneous with it, and so it is likely that both laterals and sub-laterals have evolved from the somasteroid virgalia, but which of the two columns is a homologue of the *lateral shields* of later ophiuroids is not at all clear[32]; some authorities[12] suggest that the shields are formed by the fusion of both series. The mouth frame closely resembles that of the early somasteroids. The disk skeleton consists of small imbricating scale-like plates on both sides.

The next forms to appear are *Eophiura* and *Palaeura*, both from the Upper Arenig beds. The main advances over *Pradesura* are the lateral expansion of both lateral and sub-lateral plates and the gradual change, in *Palaeura* at least, from the early arrangement in which the tube-feet bases are shared more or less equally between two adjacent ambulacrals to the more ophiuroid-like condition in which far more of each basin lies on one plate.

Next, in forms like *Stenaster* (Fig. 7*c*), we see that the ambulacrals are tending to move inwards and to oppose rather than alternate. This latter change is important to the ophiuroids, because it led to the possibility of fusion of the two members of an ambulacral pair to form the typical *vertebrae* of modern forms. There is little doubt that these fossils form a series representing the earliest experiment into an ophiuroid condition; its members are usually classified into an order, STENURIDA. In all these early ophiuroids only the disk and bases of the arms are known, probably because they were burrowers, and like some of the recent starfishes and brittle stars, sat in the substratum with only the arm-tips exposed above the surface; at death, this left the disk and arm bases more favourably placed for preservation.

In the Ordovician we find the first members of the order OEGOPHIURIDA, which may have arisen from the stenurids. In these, fusion of the opposing ambulacral plates occurs, but they still lack the dorsal and ventral plates so typical of the arms of modern brittle stars. It has recently been shown[32, 33] that the peculiar living Indonesian ophiuroid *Ophiocanops* (p. 54) is a representative of this order, so a good deal can be inferred about the soft parts of the fossil forms. The chief fossil representative of the group is the Silurian form *Lapworthura*[44] (Fig. 7*d*), and like *Ophiocanops* it probably had the gonads and gut caeca extending into the arms in an asteroid-like way. *Ophiocanops* has no respiratory bursae in its disk, so it is not surprising that it also lacks radial shields, to which the muscles operating the bursae are attached; *Lapworthura* also lacks radial shields.

The oegophiurids probably gave rise to two subsequent lines: the PHRYNOPHURIDA, in which the laterals occupy a latero-ventral position on the arms, and in which the *ventral arm plates* appear; and the OPHIURIDA, in which both *ventral* and *dorsal* arm plates are present. The phrynophiurids include the basket-stars, such as *Asteroschema* (Fig. 7*f*), in which there may be bifurcation of the arms; and the ophiurids include most of the typical modern brittle-stars. In the ophiurids, such as *Ophiothrix* (Fig. 7*g*), arm movement is almost entirely restricted to the horizontal plane (*zygospondylous* articulation), while in the basket-stars all-round movement is possible (*streptospondyly*). As one would expect, this difference is reflected in broad differences in the biology of the two groups: whereas most of the ophiurids move in or on the surface of the substratum, many phrynophiurids are capable of clinging to objects with their arms[30], so that they can climb through the fronds of algae, etc. For instance, the euryalid *Asteronyx* is said to move over beds of pennatulids (sea-pens, of the phylum Coelenterata) feeding on their polyps. During their development some euryalids are said to pass through a zygospondylous condition, suggesting that the Ophiurida are more primitive.

5

THE ECHINOIDEA

In the sea-urchins, alone among the living echinoderms, the skeletal elements form a rigid theca or *test*. Who can fail to be struck by the symmetrical beauty of these objects when cleaned by the sea and cast up on the beach? In the more primitive regular forms the smoothness of the tests is not broken by food grooves, but these may be secondarily re-formed in the more advanced irregulars. The irregulars have secondary bilateral symmetry superimposed on the basic radial plan, giving them distinct anterior and posterior ends, so that they move in one direction only, and conferring on them the consequent advantages, well exploited, in colonising habitats forbidden to the regulars.

General body plan

The test is composed of interlocking plates radiating in rows from the apex round to the mouth in the centre of the under (oral) surface. The plates of all living echinoids are in alternating double columns, ambulacra and interambulacra; in some extinct groups the number of columns of each sort range from one to twenty or so, but all modern echinoids, regular and irregular, have two of each. The ambulacral plates are distinguished by having a pair of pores through which pass the canals leading to the tube-feet. The echinoids have a *closed ambulacrum*, so that the radial water vessels and the other radial structures are all internal to the skeleton.

In the urchin immediately after metamorphosis the whole of the aboral surface is covered by an apical disk of plates, consisting

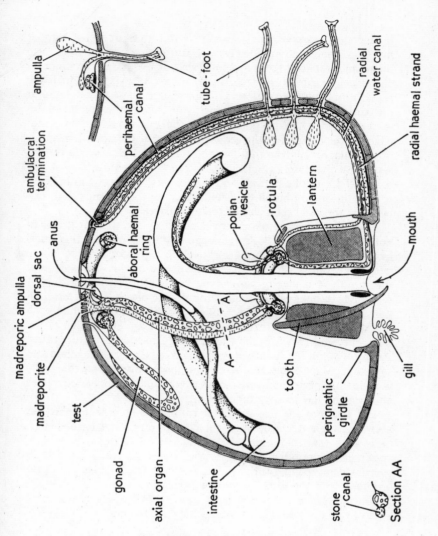

Fig. 8 Basic anatomy of an echinoid

Diagrammatic vertical section through the body of a regular echinoid, based on *Echinus*. The section is taken through one ambulacrum on the right, of which three tube-feet and their ampullae are shown. On the left is a section through the axial complex.

See Fig. 2 for key to systems.

of a central *suranal*, through which the anus opens, a ring of five *basals* (more often called *genitals*) through which the gonoducts open, and between these a second ring of five *radials* (sometimes called *oculars*, because each one contains a cavity in which the radial ambulacral structures terminate, and these places were at one time thought to be light-sensitive). One of the genital plates is perforated by many pores and is the madreporite. The outer edges of the radials are the growing points for the double columns of ambulacral plates, and the outer edges of the basals do the same for the interambulacrals. There is no equivalent of the terminal plates of asteroids and ophiuroids, unless the radials are to be considered homologous with them on the grounds that they bear the terminal tentacles of the ambulacra.

As more and more plates are budded off and the test grows in volume, the apical disk remains roughly the same size and stays in the centre of the aboral surface. Changes occur in it in the irregulars, but these are beyond the scope of this book. Round the mouth is a strong *perignathic girdle* (Fig. 8), consisting of modified ambulacral and interambulacral plates. Each terminal pair of ambulacral plates forms an arch-like *auricle*, while each interambulacral pair forms a solid *apophysis*. This girdle is the main attachment for the fascinating masticatory apparatus, *Aristotle's lantern* (described below), possessed by all regulars and some irregulars. Stretched across the space bounded by the girdle is a flexible peristomial membrane, containing embedded plates and bearing ten sensory tube-feet (p. 138).

The five gonads are attached to the inner side of the test aborally, their ducts opening through pores in the basal plates. Only in a few echinoids is it possible to determine the sex of specimens externally; in these, the males generally have their sperm ducts opening on papillae, while in the females the pores are flush with the test. Sometimes hermaphrodite individuals are produced, and a few cases have been recorded of urchins with part of one gonad male and the other part female.

Aristotle's lantern

The complex masticatory apparatus which is situated inside the mouth of every regular and every clypeasteroid sea-urchin was so called by Klein in the eighteenth century, apparently because of its similarity to a Greek lantern and the famous Greek natural historian's fascination for the sea-urchin. It consists of forty skeletal pieces, intricately interbound with muscles and connective tissue. Its purpose is to bear five strong and constantly growing teeth in such a way that they can rasp encrusting organisms, such as ectoprocts and algae, from the surface of the substratum over which the urchin is walking; to do this the teeth must be able to move up and down in relation to the mouth and to move towards and away from each other. So the teeth are held in long skeletal pieces, the *pyramids* or *alveoli*, interradial in position. Four sets of muscles pull on these: first, there are the *interpyramidals* or *comminators*, between adjacent pieces which pull the teeth together; then there are five pairs of *protractors*, a pair running from the top of each pyramid to the nearest apophyses of the perignathic girdle, which pull the teeth down towards the substratum; thirdly, there are five pairs of *retractors*, each pair originating low down on the side of each pyramid and inserting at the top of the adjacent auricles of the girdle; and lastly there are five pairs of *posturals*, continuous with the protractors but separately innervated, which move the pyramids laterally.

Two small accessory pieces lie radially between the aboral ends of the pyramids: the *rotulae* rest on top of the sides of two adjacent pyramids, and the *compasses*, rod-like with bifurcated ends, lie on top of the rotulae. Adjacent compasses are joined by the *circumferential compass muscles* which form a pentagonal pattern on the top of the lantern, and each compass has two *radial compass muscles* at its outer end, attaching it to the two apophyses to either side of it. The set of compasses, rotulae and muscles serve a function only indirectly concerned with feeding: they act as a system which pulls the perioesophageal coelomic membrane up and down so that coelomic fluid is drawn into and out of the accessory gills round the peristome which supply the muscles of the lantern with oxygen. Clypeasteroid urchins have no gills, and their peripharyngeal region is not cut off from the rest of the coelom; in consequence this set of structures is missing.

The alimentary canal

Arising from the largest plates in the peristome are ten buccal tube-feet. These are almost entirely sensory in function, and can be seen exploring the substratum over which the urchin is moving. If it passes over encrusting organisms, the teeth are set working to rasp off the food. This then passes into an *oesophagus*, which ascends up the centre of the lantern and on to its aboral surface,

C

where there is considerable enlargement into an *intestine*, running right round the inside of the body to its starting-point, doubling back on itself and running halfway round again, then ascending to the anus. A curious structure associated with the gut, the *siphon*, is present in echinoids, but does not seem to be found in other echinoderms. This is a tube which comes off the gut at the start of the intestine, runs along parallel to it until the point of doubling, then re-énters it. Its function is not certain; there is said to be a current in it, so it may be for removing water during digestion. In some irregular echinoids there are reported to be *two* siphons, each bypassing a different region of the gut. There is an interesting developmental feature of the mesenteries holding the main folds of the intestine: they are horizontal, but start life as the vertical sheet of tissue between the two posterior coelomic sacs of the embryo (p. 150). The urchin's adult mouth is formed on one side of the larva, so this sheet of tissue becomes horizontal in respect of adult orientation.

The coelom

Students who crack open an echinoid test are usually disappointed to find that most of the inside is a fluid-filled cavity, with the gut hugging the inside wall of the test and the axial complex hanging from the apical disk in the middle like a piece of black string. Certainly, the perivisceral coelom is more extensive than in any other class. It is also considerably subdivided, there being two distinct cavities round the anus, a *perianal* coelom, and outside that a *periproctal* coelom. Then, surrounding the whole of the lantern, as already mentioned, is the *peri-oesophageal* coelom. The large volume of the main part of the perivisceral coelom is probably wholly accounted for by the size to which the gonads swell in season.

The tubular coelomic and haemal systems

The circum-oral ring vessels of these systems lie just above the lantern (Fig. 8), and the radial water vessels pass under the rotulae, down the sides of the lantern adjacent to the alveolar muscles and under the arches of the auricles before passing up the inside mid-line of the ambulacra, while the radial haemal strands pass down the *inside* of the alveoli of the lantern, then between adjacent alveoli before passing under the arches of the auricles. The various radial structures bear the same relationship to each other in the ambulacra as they do in the ophiuroids; that is, the epineural sinus lies between the plates and the radial nerve cord, the perihaemal (=hyponeural) sinus comes next, then the haemal strand and lastly, most internal, the radial water vessel. The oral haemal ring gives off a branch leading to the very extensive lacunar system in the gut walls, and there is an aboral haemal ring, surrounded by a perihaemal space, giving off branches to the gonads. One curious thing about the perihaemal system is that the circum-oral ring does not appear to be in communication with the radial perihaemal canals. An axial sinus is absent[47].

The axial organ

This organ, which is the axial part of the haemal system, lies adjacent to the stone canal in the strand of tissue, usually called the axial complex, which runs from the top of Aristotle's lantern to the aboral pole of the animal. It consists[58] of a mass of spongy haemal tissue which has *embayments* and *canals* permeating it. Some of these spaces are confluent with a *central cavity* in which a separate *pulsating vessel* lies. The pulsating vessel sends out branches into the surrounding tissue, and in places the wall between these branches and the embayments and canals is pierced by *funnels*. So all the spaces in the haemal tissue are open to one another, though which cavity actually leads off to the rest of the haemal system is not clear.

The organ has been shown to have several functions. First, substances are present in it, such as tryptophan, which are known to be constituents of some hormones, so the organ may in part be endocrinal; secondly, the organ plays a part in ridding the body of foreign material. At times, and particularly following the introduction of extraneous cells or other foreign matter, the embayments and canals are seen to have cyst-like masses of amoebocytes within them, being whirled around and wrapped in a mucous envelope, the mucus probably being pumped around the cavities in the organ by the pulsating vessel. Later, the cysts are extruded from the canals into the perivisceral coelom, but what becomes of them subsequently is not clear. Thirdly, the walls of the haemal spaces, including those of the pulsating vessel, proliferate epithelial cells, particularly during tissue repair following injury to any part of the test, and it appears that the cells migrate to the injured area where they help to form new tissue.

The nervous system

The main nerve ring is situated round the oesophagus inside the pyramids of the lantern, and the radial nerves leave it just below the radial haemal strands, following their path between the pyramids, under the auricles of the perignathic girdle and out to the ambulacra. Because these main nerve tracts are beneath the skeleton, they are somewhat isolated from the main sensory basi-epithelial plexus which covers the whole body, being joined to it only by tracts passing out of the ambulacral tube-foot pores.

Apart from the general sensitivity of the basi-epithelial plexus, there are few organs of special sense. The balance organs, *sphaeridia*, are dealt with in Chapter 11. Only in the family Diadematidae, apparently, are there any special light-sensitive areas.

The evolution and adaptive radiation of the echinoids

There is no clear-cut evidence as to the origin of the echinoids. Time-relations do not allow their derivation from the Ophiocistioidea of the Silurian (p. 106), the only other echinoderm group with any sign of a masticatory apparatus, since the first echinoid, *Bothriocidaris* (Fig. 9a), occurs in the Ordovician. Mortensen, the greatest authority on the echinoids, whose *Monograph of the Echinoidea*[59] is a monument to a life's study of the group, thinks that the echinoids arose from close to the *Stromatocystites*-like edrioasteroids (Fig. 17c) of the Cambrian, on the grounds of similarity of plate arrangement. As in the asteroids, such a derivation would mean principally an inversion, so that the mouth came to face downwards. It is not easy to visualise how this rather drastic change of habit might have occurred—it is hard to see it as anything but a sudden step, which would mean a correspondingly sudden change in feeding habits. Admittedly, if the edrioasteroid stock was ciliary-feeding (which would seem to have been the case) this method could still have been effective if the animal inverted, particularly if selection acted to extend the ambulacra on to the aboral (now upper) surface, as indeed did happen in some of the later edrioasteroids. Then, advantage would probably be gained by converting some of the peristomial plates into a dental apparatus for scraping organisms from the substratum.

So it is not surprising that even in the oldest known echinoids we see a primitive masticatory apparatus, forerunner of the complex Aristotle's lantern. Neither is it entirely unexpected that the tests of the earliest echinoids show signs of having been flexible, as were some edrioasteroids.

Though *Bothriocidaris* is the oldest known echinoid, it is not

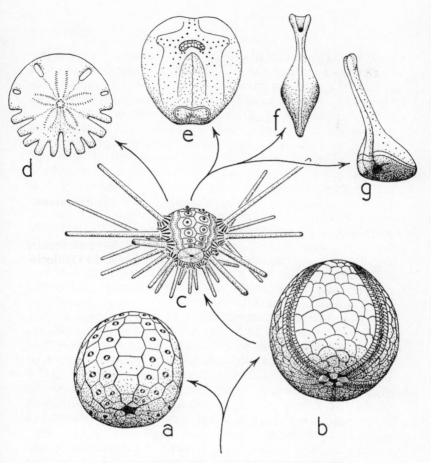

Fig. 9   Evolution and adaptive radiation in the echinoids

a *Bothriocidaris* (Ord.), the earliest echinoid, with three columns of
plates in each ambulacrum, two bearing the pore-pairs and a
central column devoid of tube-feet but bearing the radial canal
internally.
b *Aulechinus* (slightly later Ord.), regarded as the echinoid closest to
the ancestral type. Many plates in each interambulacrum.
c *Cidaris*, a typical regular echinoid, with two ambulacral and two
interambulacral columns. Spines of the nearest interambulacrum
removed.

d to g, examples of recent and fossil irregular echinoids, spines removed.
d *Rotula*, a clypeasteroid sand-dollar, aboral view, showing lunules.
e *Spatangus*, a typical spatangoid heart-urchin, oral view.
f *Echinosigra*, a pourtalesiid holasteroid, oral view.
g *Hagenowia*, a holasteroid from the Cretaceous, in which a rostrum is
present.

nowadays regarded as ancestral[49] because there are great differences between it and the early members of the next group to appear, the ECHINOCYSTITOIDA, from very slightly later in the Ordovician. The test of *Bothriocidaris* was probably fairly rigid, with three columns of plates in each ambulacral area, and no interambulacral columns as such. The radial water vascular canal, remains of which have been found preserved in some of the fossils, ran within the central column of each set of three, and gave off branches to tube-feet which emerged through pores in the other two columns of the set. A further surprising freak of preservation has occurred: the recognisable imprints of tube-feet have been found associated with the pore-pairs of some ambulacral plates. In the peristomial region are five plates, believed to be the forerunners of teeth. In contrast, early members of the slightly later echinocystitoids, such as *Aulechinus* (Fig. 9b) and *Palaechinus*, usually had a slightly flexible test formed of two columns of ambulacral and an indefinite number of interambulacral plates. From this, we infer that the first echinoid also had two columns of ambulacrals and probably an indefinite number of interambulacrals, which would fit in nicely with the theory of edrioasteroid ancestry. There is another feature in forms such as *Aulechinus* which suggests an edrioasteroid ancestry: they have an open ambulacrum, the radial water canal lying in a groove on the outside of the ambulacral plates, not in the interior of the test. Closure of the ambulacrum, apart from that seen in the separate bothriocidaroid line, occurred during the evolution of the echinocystitoids.

After the establishment of the primitive echinoid there was, then, early specialisation to the somewhat inflexible bothriocidaroid condition, which left no descendants, and, in the 'main' line, a time of experimentation from the Silurian to the top of the Permian in which various patterns of plate arrangement and methods of strengthening the test were tried. So we see a trend in some forms, such as *Melonechinus* (Carboniferous), towards a large number of ambulacral columns, and in others, such as *Lepidocentrus* (Carboniferous), to a large number of interambulacra. In none of these early forms is there a perignathic girdle for a strong attachment of the masticatory apparatus.

Some time in the Silurian one group of echinoids, the CIDAROIDA (Fig. 9c), mostly restricted the number of columns in each ambulacrum and interambulacrum to two, and the group kept this pattern steadily for the rest of the Palaeozoic; they also evolved a perignathic girdle. Then, at the end of the Permian, something remarkable happened to the environment, something which probably happened the world over and which had the effect of apparently wiping out every echinoid genus except one, belonging to the cidaroids, called *Miocidaris*. Why this was the only one to survive nobody can say with any certainty, but it alone entered the Mesozoic era to provide the stock from which all subsequent echinoids arose. By Upper Triassic times regular echinoids other than cidaroids were also appearing, and there seems to have been a rapid deployment in the Jurassic, culminating at the present day in some sixteen families[49]. Most of these live on rocky bottoms, feeding on surface detritus and organisms which can be chiselled off by the teeth, and protecting themselves by means of spines and pedicellariae, many of which are poisonous. Several genera of various families, however, such as *Paracentrotus* of the Echinidae, may excavate cavities for themselves in the rock, in order to gain protection from wave-action. They do this by clinging to the rock-surface by their tube-feet and continuously abrading the rock with their spines and teeth[61]. They feed either on particles which are washed to the bottom of the cavity, or they may turn in the cavity to bring the mouth uppermost. The interesting point about these boring species is that some populations of apparently the same species do not bore, even though a similar type of rock may be available to them. For instance, *Paracentrotus lividus* on the coast of Ireland and North-west France may honeycomb the rock with its cavities, while members of the same species in the Mediterranean hardly ever bore into the rock.

At about the beginning of the Jurassic we see the first signs of a break in the almost perfect radial symmetry of the urchins, a step which gave them definite front and back ends and which opened up new habitats hitherto forbidden to the group. There appear to have been two main attempts at irregularity: one, grouped as the GNATHOSTOMATA, retained the Aristotle's lantern and girdle, whilst the other, the ATELOSTOMATA, dispensed with it. The earliest irregular urchin to appear in the record is the atelostome *Pygomalus*, from the Sinemurian stage of the Lower Jurassic; the earliest gnathostomes were *Holectypus* and *Galeropygus* from the slightly later Pliensbachian. Both gnathostomes and atelostomes persist to the present day, the gnathostomes containing two orders, HOLECTYPOIDA and CLYPEASTEROIDA, and the atelostomes being represented by the CASSIDULOIDA, HOLASTEROIDA and SPATANGOIDA (Fig. 9e, f, g). Though there can be little doubt of their separate derivation, the gnathostomes and atelostomes show many parallel changes, the chief one being the marked specialisation of tube-feet on various parts of the animal for particular jobs: in both, for instance, we see the evolution of special areas of non-extensile tube-feet on the dorsal surface for respiration only, and in some members of both we see the complete loss of suckered tube-feet. Both these trends are associated with the adoption of a burrowing habit, with its consequent protective advantages, a habit exploited to the full by the spatangoids. Among the gnathostomes the clypeasteroid sand-dollars achieve probably the greatest specialisation, some, such as the Key-hole Urchin, *Rotula* (Fig. 9d), becoming remarkably flat and possessing holes through the test, from the upper surface to the lower, probably to aid burrowing. They sit in the substratum either flat and just below the surface, rather like a flatfish, or at an angle with an edge of the test protruding to face oncoming currents, presumably as a food-catching device. The tiny British urchin *Echinocyamus*, sometimes found inshore on shell-gravel or coarse sand but more often offshore, is a clypeasteroid. But it is not a sand-dollar; these flat forms do not occur in Britain.

The spatangoids (Fig. 9e) are far more ingenious in their mode of burrowing. Some of them, such as the common British heart-urchin of sandy shores, *Echinocardium*, have tube-feet on different parts of the body specialised for burrow-building and feeding (p. 142) and they can burrow at depths of anything up to five times their own height, all the time maintaining the vital connexion with the surface of the substratum for food and oxygenated water[60]. The mouth in spatangoids is usually near the front of the underside, and the anus usually on the posterior surface. Feeding is by a combination of ciliary action and the activity of sticky tube-feet, which pass material from the substratum into the gut where the organisms adhering to it are digested off; the particles of the substratum are then voided from the anus. The most peculiar atelostomes are probably the deep-water pourtalesiids[56], such as *Echinosigra* (Fig. 9f), in which the mouth is at the very front and the anus almost at the rear, giving it a strong resemblance to a holothuroid in shape. Another curious form was the Cretaceous *Hagenowia* (Fig. 9g) in which the aboral surface is drawn out into a rostrum.

6

THE HOLOTHUROIDEA

The sea-cucumbers are eleutherozoan echinoderms that lie on one side with the mouth and anus at the opposite ends of a cylindrical body. By no stretch of the imagination can most of them be described as elegant: they are limp and slimy to the touch, and some exude special sticky strings to entangle anything (and anyone) who meddles with them. Their skins are usually tough and leathery, with embedded spicules. Yet the Chinese relish a meal of *trepang*, which consists of sun-dried body walls of several species of *Holothuria, Stichopus* and *Thelenota*. 'It is employed,' says Forbes, '. . . (in) the preparation of nutritious soups, in common with an esculent sea-weed, sharks' fins, edible bird's-nests . . . affording much jelly.' Certainly, trepang appears to be highly nutritious, and, as its protein constituents are almost completely broken down by pepsin, highly digestible.

Holothuroids feed by means of the buccal tube-feet. Either they move slowly over the sea-bottom, sweeping the area ahead or under them with their sticky arborescent tentacles, or they nestle in nooks in the rock, and sweep the 'forecourt' for settling organisms, or they burrow into the soft substratum and sit with their buccal tube-feet forming a collecting net for the rain of detritus. A few holothuroids are pelagic, e.g. the elasipod *Pelagothuria* (Fig. 11g), in which parts of the body wall and tube-feet are extended to form a swimming membrane; they apparently feed on plankton. These forms are much more beautiful than the others, and closely parallel the jellyfishes in morphology and ecology.

General body plan

The ambulacra pass down the body lengthwise (Fig. 10). They are arranged so that three ambulacra lie ventrally (the *trivium*) and two dorsally (the *bivium*). When tube-feet are present along the body, that is, in all groups except the Apoda, those of the trivium are generally suckered, while those of the bivium are papillate. Often, as in *Holothuria*, the tube-feet arise all over the body, not restricted to ambulacra, so that the trivium forms a creeping sole and the bivium a dorsal sheet of mainly sensory papillae. Only one gonad is present, in the mid-dorsal interambulacrum, opening by a gonopore just behind the mouth. In those in which there is a madreporite this too is anterior.

It is at the posterior end of the alimentary canal that the main differences between these and other echinoderms are found; here, special respiratory structures and, in some, the special defensive organs are present as outgrowths of the cloaca, to be explained later. They occupy a large part of the general body cavity.

On the outside of the body there are no spines, no cilia and no pedicellariae. The body wall itself contains the usual calcareous elements but in this case they are reduced to spicules in the shape of rods, crosses, hooks, anchors or wheels. One worker has estimated that there may be as many as 21 million spicules in one individual, supporting the outer body wall, the tube-feet and some internal tissues. Sometimes they form a protective armour, like roof tiles. The shape and combination of these spicules are used extensively by taxonomists. It would be interesting to know how each shape fits into the pattern of activity of the soft tissues in which they are contained, and how specific differences in spicule shape are correlated with differences in ecology and behaviour between close relatives; but this problem of functional taxonomy would be a very difficult and challenging one to tackle.

As in asteroids and ophiuroids, there is a complete ring of much larger ossicles round the mouth and oesophagus, to which muscles of the buccal tube-feet and body wall are attached. When the feeding animal is disturbed the mouth and buccal tube-feet can be withdrawn into the body on an introvert, by contraction of longitudinal muscles of the body wall inserted at the calcareous ring. Extension is brought about by contraction of circular muscles, acting on the coelomic fluid. As one would expect from animals inhabiting a variety of ecological niches, the buccal tube-feet show a wide range of form. To some extent this follows the

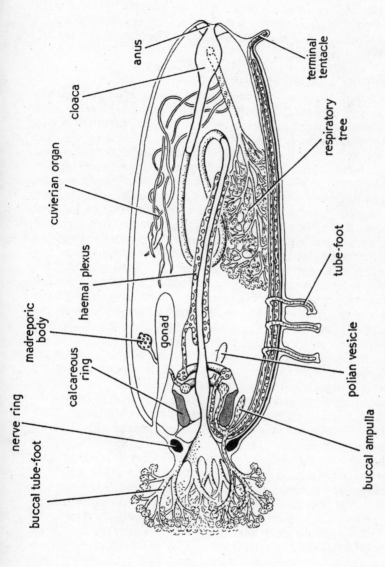

Fig. 10 Basic anatomy of a holothuroid

Diagrammatic vertical section through the body of a generalised holothuroid, based on *Holothuria*. The section is taken through the ventral ambulacrum, of which three tube-feet are shown. In the posterior region only one of the pair of respiratory trees and cuvierian organs is shown.

See Fig. 2 for key to systems.

division of the holothuroid orders, classified on other grounds. In the Dendrochirotacea the buccal tube-feet are mostly dendritic; in the Dactylochirotacea they are digitate; in the Aspidochirotacea they terminate in a circular disk; and in the Apodacea they are digitate or pinnate.

Alimentary canal and its accessory structures

The mouth leads into an oesophagus which passes through the calcareous ring to a slightly enlarged part, the stomach, and thence to an intestine whose loops follow the same pattern in relation to the orientation of the ambulacra as those of crinoids, though drawn out longitudinally. The intestine ends in a cloaca, and it is from here that the accessory structures arise, when they are present. A pair of these structures, the *respiratory trees* (Fig. 10), arises from the cloaca of many holothuroids. They are arborescent tubules, lined with the same layers as the gut and ending in thin-walled vesicles. The cloaca, pumping at a rate of six to ten pulses per minute, and working in conjunction with an anal sphincter, forces water into the lumina of the trees, so that gaseous exchange takes place between the sea-water and the coelomic fluid. A curious example of animal relationship occurs here: there is a teleost fish, *Carapus*, which lives in the main stems of the trees of aspidochirotes. Its head protrudes from the anus and it uses its host apparently solely for shelter, catching its own crustacean food during night sorties.

In aspidochirotes there are two groups of special defensive *cuvierian organs* branching from the bases of the respiratory trees (Fig. 10). These are long structures, sometimes hollow, with an outer layer of special cells, two layers of muscle fibres just below the surface, and a thick layer of collagenous connective tissue. When the animal is irritated the structures are extruded steadily from the anus and acquire a sticky surface, thus trapping the aggressor. The British species *Holothuria forskäli* will perform in this way when handled, and the mess one individual can make must be seen to be believed. The process has given rise to the common name 'cotton-spinner' for *Holothuria*. The mechanism of extrusion is believed[65] to have two components: first, water from the lumen of the respiratory trees is forced into the tubules, the proximal ends of the tubules are shut off by muscles, and the muscles in the walls of the tubules contract, elongating them by hydrostatic pressure. The cloacal wall then splits at a weak point formed by the crossover of circular and longitudinal muscles, so that the tubules are forced out through the anus to elongate steadily into threads sometimes several feet long. The elongation splits the covering cells, freeing a sticky substance, and the collagen fibres keep the threads from breaking, thus making a formidable trap for the unwary. The process has its parallel in the holothuroids without cuvierian organs: these forms when annoyed split the intestinal wall near the anus but simply spew out the gut and respiratory trees, to regenerate them again later. The significance of this operation in the life of the animal is not clear, but possibly the gut and associated organs provide a tasty meal for a predator bent on having one, leaving the tougher remains of the animal to creep quietly away.

The coelom

This is subdivided by mesenteries into four cavities: the main perivisceral part, an anterior perioesophageal part with a peribuccal part within it, and a posterior perianal part. The gut is held in the coelom by mesenteries, and in the Apoda these contain special *ciliated urns*, which apparently collect waste matter enclosed in coelomocytes. What happens to the waste subsequently is not clear, but some authorities say it is passed out through the body wall.

The tubular coelomic and haemal systems

Only the water vascular and haemal systems are represented to any extent, though a channel corresponding in position to the perihaemal canal has been described in the ambulacra of some holothuroids. There is normally no axial complex as such, and though the stone canal is always present, it very often does not open at an exterior madreporite but terminates within the coelom at a *madreporic body*, a swelling with many ciliated pores piercing it. The water ring lies posterior to the ring of ossicles (Fig. 10) and gives off a number of polian vesicles (one in *Holothuria forskäli*, up to thirty in some others) as well as the five radial canals. The canals pass forwards, down the sides of the mesentery forming the wall of the perioesophageal coelom, give off the lumina of the buccal tube-feet and their ampullae and continue inside the body to the posterior end, where they terminate as the lumina of the terminal tentacles. The order Apodida is an exception to this: in it there are no radial coelomic vessels at all, only the circum-oral ring vessels and buccal tube-feet.

The haemal system is well developed, mostly in association with the gut, where there are two main branches on either side of it, joined by a mass of capillaries, a *rete mirabile* ('wondrous network'), over the surface of the gut wall. The oral haemal ring, aboral to the water vascular ring, gives off the usual five radial haemal strands, lying with the same relationship to the radial elements as in other echinoderms. The axial organ is probably absent, though some say it is represented by a small knot of tissue arising from the haemal ring adjacent to the origin of the stone canal.

The nervous system

The main nerve ring, as in other echinoderms, lies nearer the mouth than do the coelomic ring components, and is embedded in the peristomial membrane close to the bases of the buccal tube-feet. It sends radial branches out under the radial components of the calcareous ring, and these branches ramify into the general basi-epithelial plexus of the body wall, continuing into the tube-feet. The system just described is the *ectoneural system*. But the radial nerve cords are not single components: a section through the ambulacrum of a holothuroid shows that in addition to the ectoneural part there are aggregations of neurones internal to it and to some extent ramifying with it. This is the *hyponeural system*, homologous with that of asteroids (Lange's centres), and possibly also with the deep oral system of crinoids. It is mainly motor, but it too receives fibres from the basi-epithelial plexus of the body wall.

The main sense organs of holothuroids are the tactile and chemosensory cells scattered in the epithelium, and, in some apodans, aggregations of nerve cells in special warts on the body surface. In some aspidochirotes and apodans there are, in addition, statocysts embedded in the body wall close to the point where the radial nerves leave the nerve ring. These are tiny fluid-filled spheres containing calcareous statoliths, the differential movement of which is registered in special nerves and conveyed to the radial nerve, presumably to right the animal if its orientation is upset by external forces.

The ambulacrum

From what has been said about the radial components of the various systems in the ambulacra it will be evident that holothuroids, in common with ophiuroids and echinoids, possess a *closed ambulacrum*. Like them, the enclosure of the radial elements inside the body wall can be followed during development, and results in a small canal, the epineural sinus, remaining between the body wall and the radial nerve. We have seen that in each of the classes with a closed ambulacrum this sinus is present, as is an equivalent one external to the nerve ring round the mouth. Whether this sinus communicates with the coelomic spaces is not known, but one must conclude that for physiological reasons the nerve cords must be bathed on either side by fluid, on the inside by that in the radial perihaemal sinus and on the outside either by the sea-water (open system) or by the fluid in the epineural sinus (closed system).

Evolution and adaptive radiation of the holothuroids

It will be recalled (p. 68) that primitive echinoids from the Ordovician often had flexible tests. In fact, the echinoid *Eothuria* was mistakenly taken to be a fossil holothuroid by its original describers, as its name suggests. So by this evidence and the added fact that some present-day holothuroids have imbricating plates investing their bodies, it is possible that the echinoids and the holothuroids are related. Some palaeontologists[12] would have it that the cigar-shaped helicoplacoids (p. 105 and Fig. 18*a*) are in some way connected with the holothuroids, but these Lower Cambrian echinoderms do not show pentamerous symmetry, and if present ideas on the significance of such a body pattern are correct, it is not easy to see pentamery arise except in a rigidly plated animal. Others see a similarity between the perioesophageal calcareous ossicles of holothuroids and the plating on the oral surface of edrioasteroids such as *Isorphus* (Upper Ordovician) and in this case the inward migration of the plates is seen as a consequence of the development of an introvert requiring a rigid muscle attachment. But such a derivation is far from certain.

Does the fossil record help? Decidedly not. The great majority of holothuroid remains are, of course, only the ossicles, and the relationships of these, even in Recent forms, are not clearly defined. The earliest occurrence is of an animal given the name *Thuroholia* from the Ordovician, and then in the Devonian is an impression of a whole animal, named *Palaeocucumaria*. Further impressions of whole animals turn up in the Solnhofen Slates of the German Jurassic, but these give away so little that they hardly enter the picture.

So our ideas on the evolution of this class are gained almost solely by a study of the comparative anatomy of Recent forms. On this basis, it is assumed that the early members of the class had plated bodies, like today's *Placothuria* or *Psolus*; the plates of *Thuroholia* (Ordovician) resemble those that invest the body in some present-day forms. Then from such a form three lines are thought to have diverged: one led to the surface-dwelling Aspidochirotacea, by reduction of the plates to mere ossicles; another to the Dendrochirotacea, some of which retained the body-plating; and a third to the Apodacea, which reduced both the plates to ossicles and the water vascular system to the oral tube-foot system only.

The holothuroids show a surprising degree of adaptive radiation. Some members are benthic, some pelagic and some burrowing; they are found in littoral waters or at great depths; some are exceedingly sluggish, while others exhibit a fair turn of speed when necessary.

The benthic, creeping holothuroids are mainly but not entirely included in the orders DENDROCHIROTIDA and ASPIDOCHIROTIDA. The dendrochirote *Psolus* (Fig. 11a) has its ventral trivium flattened to form a creeping, muscular sole for locomotion, with very few tube-feet involved, rather like the foot of a mollusc, while the aspidochirote *Holothuria* has tube-feet over the whole ventral surface, and these are the locomotory organs. The mouth in these creeping forms may be at the anterior end, so that the oral tube-feet can sweep the sea-floor for food, as in some species of *Cucumaria* (Dendrochirotida) and *Holothuria* (Aspidochirotida) or it may point dorsally, as in *Psolus*, probably to collect falling detritus. Members of the other order with tube-feet present down the length of the body, the ELASIPODIDA, show a variety of structures forming sails and other flotation devices: these forms are mainly bathypelagic, and are known mainly from the collections of major expeditions[70]. One of the most curious is *Pelagothuria* (Fig. 11g), known principally from collections made by the *Albatross* and other vessels, in which parts of the body wall are extended as radially arranged papillae supporting a web[63]. The mouth, surrounded by oral tube-feet, is directed upwards and the anus downwards, so that the form somewhat resembles an inverted medusa. Some species have been taken at great depths,

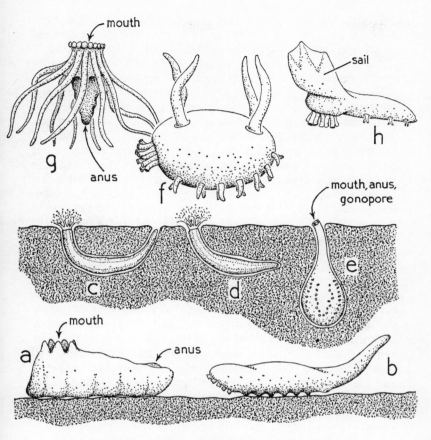

Fig. 11 Adaptive radiation in the holothuroids

a *Psolus*, a dendrochirote, with mouth and anus directed upwards.
b *Psychropotes*, an elasipod, with mouth and anus directed downwards.

c to e, some burrowing holothuroids.
c *Cucumaria*, a dendrochirote which uses anal respiration.
d *Synapta*, an apodan, which lacks respiratory trees, and does not
    appear to need a U-shaped burrow.
e *Rhopalodina*, an unusual dactylochirote, in which mouth, anus and
    gonopore all open at the end of a spout.

f *Scotoplanes*, a dendrochirote with large dorsal papillae.
g *Pelagothuria*, a pelagic elasipod.
h *Peniagone*, an elasipod with part of the body wall raised as a sail.

others at moderate depths and others actually at the surface. Another form is *Peniagone* (Fig. 11*h*), with the mouth directed downwards and part of the dorsal body wall raised into a sail.

Burrowing holothuroids are found mainly in the order Dendrochirotida and the two orders lacking tube-feet down the sides of the body, the MOLPADIIDA and APODIDA. Of these three, the latter alone lacks respiratory trees, so its members are not restricted to U-shaped burrows, as are members of the other two. The dendrochirote *Thyone* burrows by wriggling the middle of the body, so that this part alone sinks below the surface, while *Cucumaria*, in the same order, goes down head-first until it is buried, then moves round and upwards to break surface a few inches in front of the original hole. It then extends its oral tube-feet to catch food[66]. If the original hole falls in, the tail will be pushed up to the surface again (Fig. 11*c*), so that water can be taken into the respiratory trees through the cloaca. The burrowing molpadiids, such as *Molpadia* and *Caudina*, also require two openings to the burrow, and these forms usually possess a thin caudal region which can be pushed to the surface of the substratum for respiration and defaecation. It is interesting in this connexion that in the British dendrochirote *Cucumaria elongata* the faeces are voided with some force, so that they do not drop back into the burrow to foul the water used for respiration. The apodans, such as *Synapta*, do not possess respiratory trees, and though they too push themselves into the substratum head-first and move round so that their head protrudes from the surface, they do not need to keep the anus in contact with the water (Fig. 11*d*): respiration in this case is probably solely by diffusion across the walls of the oral tube-feet.

7

EXTINCT CRINOZOANS

CYSTOIDEA
EOCRINOIDEA
PARACRINOIDEA
PARABLASTOIDEA
BLASTOIDEA

In the next three chapters we shall consider those groups of the echinoderms which are entirely extinct. The cystoids and blastoids, to be dealt with first, are groups of brachiole-bearing echinoderms which probably left no descendants. They may be related, because the thecal pores of the two groups have similarities; but until more is known about them and about associated minor groups, such as eocrinoids, paracrinoids and parablastoids, the phyletic picture cannot be properly assessed.

THE CYSTOIDEA

The class Cystoidea, once a 'rag bag' containing almost all the extinct forms that could definitely be assigned to the phylum, contains those early echinoderms whose plates are pierced by special pores which in life bore structures apparently independent of any internal system of coelomic canals, such as a water-vascular system. The *theca* (Fig. 13*a*) is flask-shaped with the *mouth* in the centre of the upper surface and the *anus* to one side of the mouth, surrounded by a circlet of five or six plates. Between the mouth and anus are usually two other pores, probably a *gonopore* (nearest the anus) and a *hydropore*. The presence of a hydropore seems to indicate that there was a water vascular system, though no other trace of such a system has yet been found. Some cystoids, particularly the later forms (Silurian and Devonian) were borne on a stem[85], which in life may have adhered to foreign objects on the sea floor; but more often the theca was itself attached in some way to a hard object. The food-collecting apparatus consisted of a number of grooved *brachioles*, arising either close to the mouth or from the sides of ambulacral grooves on the surface of the theca. The thecal pores, unique to the group, and generally regarded as having been respiratory organs, have formed the basis for the accepted classification into two orders, Rhombifera and Diploporita. Within these two orders one can trace two evolutionary trends in parallel: a gradual increase in efficiency of the respiratory organs, and a lengthening of the food grooves on the theca. With new finds and improved techniques in recent years, it has become clearer how the respiratory organs of the two orders may have functioned.

Dipores

As their name suggests, these normally pierce the theca in pairs. In surface view (Fig. 12*a*) a single dipore consists of an oval

Fig. 12   Structure of cystoid pore systems

a to *d*, diploporite dipores.
a Surface view of a dipore in a fossil diploporite.
b Section across the fossil dipore.
c Tentative reconstruction of a dipore having only a thin integument across its depression.
d Tentative reconstruction of another type of dipore with a thin skeletal sheet within the respiratory integument.

e A rhombiferan cryptorhomb with simple inhalent pores and a funnel-like exhalent aperture.
f A fistulipore.

g to *j*, tentative reconstructions of various types of rhombiferan pectinirhombs.
g A simple folded-plate system, as seen in *Macrocystella*.
h A conjunct pectinirhomb with discrete dichopores, as in *Cheirocrinus*.
i A conjunct pectinirhomb with confluent dichopores, as in *Pleuro-cystites*.
j A disjunct pectinirhomb with confluent dichopores, as in *Echinoen-crinites*.
(*c* to *j* based on Paul [74])

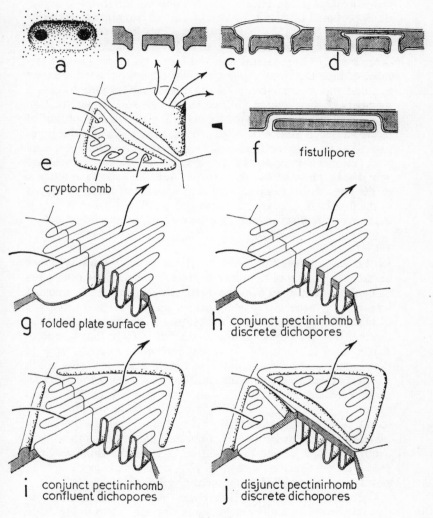

a

b

c

d

e cryptorhomb

f fistulipore

g folded plate surface

h conjunct pectinirhomb
discrete dichopores

i conjunct pectinirhomb
confluent dichopores

j disjunct pectinirhomb
discrete dichopores

Fig. 12

depression with the two pores near each end. Each pore passes
right through the plate, and the opening on the inside is sometimes bordered by tiny projections from the plate into the interior
of the body (Fig. 12b). No cystoid has ever been described with
remains of any of its soft parts, so any reconstruction of these or
other cystoid pore systems must be based on a reasoned assess-
ment of how the living structure might have functioned, bearing
in mind the sort of structural organisation possessed by living
members of the phylum. On this basis, the most likely structure
of a dipore was similar to a papula (p. 36), in which a 'blister' of
tissue stretched across the thecal depression enclosing the pore-
pair so that coelomic fluid could circulate out of one pore and
back into the interior through the other. Gaseous exchange
would take place across the papular wall. Sometimes the tissue
of the papula was calcified, and thus capable of being preserved
in the fossil; in this case the pores and canals can be seen only if
the thin calcite sheet covering the depression has weathered away.

Rhombopores

In the Rhombifera the structure of the pore systems is rather
different from that of the dipores. Here, instead of the pore-pairs
being dotted about all over the thecal surface, they are aggregated
into groups, each group lying across an interplate suture. The
distance between the pores in the centre of a group is greater than
that between the pores at either end (Fig. 12d), so that a rhomb-
shaped area is produced. Two fundamentally different types of
pore systems occur[74]: in the first, called *dichopores* (Fig. 12d),
the pores open to the exterior and the canal joining them, sometimes calcified and sometimes not, is internal to the thecal
plates; in the second, called *fistulipores* (Fig. 12d), the pores open
to the interior, and the canal joining them is within the thecal
plate, just below its surface. Dichopores open either as slits in the
external surface of the plates, in which case the rhomb-shaped
area is called a *pectinirhomb* (Fig. 12h), or as simple or compound
pores, when it is called a *cryptorhomb* (Fig. 12e).

Clearly, the structural differences between dichopores and
fistulipores reflect a different mode of operation. Assuming again
a respiratory function, in dichopores it seems most likely that the
sea-water was pumped into the canal through one pore and out
again through the other, whereas in fistulipores it was the coelomic
fluid that was pumped through the canal system.

It is possible to take the interpretation of pectinirhomb and cryptorhomb operation still further: incurrent and excurrent parts of the structure can be recognised by differences in the pores and their configuration. For instance, in some pectinirhombs the slits are longer and narrower at what is interpreted as the inhalent side, so that foreign matter passing in with the current can always be passed out again; in others, the exhalent slits are surrounded by walls of calcite that direct the excurrent away from the rhomb to avoid recirculation. Again, in some cryptorhombs the inhalent pores are compound, like sieves, to keep out foreign material above a certain size, while the exhalent pores are simple; and in others, the exhalent pores open into a sort of chimney which is directed away from the area to prevent recirculation. If, on this basis, the currents of all pore-rhombs on a single cystoid are plotted, it appears that they flow from the two poles of the theca towards its 'equator'. Thus, on the one hand, de-oxygenated and filtered water is kept away from the feeding currents, and, on the other, currents which might disturb the sediment are minimised.

In the case of fistulipores such detailed interpretation is impossible. In accepting their respiratory function, one must assume that gaseous exchange could take place across a thin calcite sheet. Echinoderm skeletal plates are built in the form of a crystal network (see p. 123), so it is quite possible for gases to diffuse between the skeletal struts. And there is a precedent for respiration having occurred across thin skeleton-supported tissue: the hydrospire system of blastoids (p. 96) is almost universally regarded as having been respiratory.

What can be said about the evolution of such structures? It seems likely that some early cystoids respired across the entire theca. In one very early cystoid, *Macrocystella* (Lower Ordovician), the theca is thin and much folded (Fig. 12*g*), probably to improve the respiratory surface, and the folds cross the interplate sutures in the same pattern as the pore-rhombs. It is a short haul from this situation to that of *Cheirocrinus* (slightly later in the Ordovician), in which the internal parts of the folds are thin but the parts at the surface are very much thicker (Fig. 12*h*). This situation represents a *conjunct* pectinirhomb with *discrete* dichopores, and the next stage (*confluent* dichopores) is marked by a return to the folded condition, but with a strengthening rim of calcite round the whole area, as in *Pleurocystites* (Fig. 12*i*). Next, the inhalent and exhalent pores are separated (*disjunct* pectinirhomb), to improve the flow of respiratory water, and here again the pores may be discrete, as in *Echinoencrinites* (Fig. 12*j*), or confluent, as in *Callocystites*.

Much less can be said about the evolution of fistulipores, though theoretically if one starts from a simple situation like *Macrocystella* one can envisage a gradual elaboration of the currents of coelomic fluid on the *inside* of the thecal folds, until a condition like that in *Echinosphaerites* (Fig. 12*f*) is reached.

The evolution of the cystoids

Nothing can be said with certainty about the origin of this Class. It appears quite likely that the two orders, Rhombifera and Diploporita, were independently derived from a crinozoan ancestor[73], such as a primitive eocrinoid, but the steps by which this may have happened in either line are unknown.

Though there are unreliable accounts of cystoids from the Cambrian, it is probably true to say that the Ordovican saw the establishment and main deployment of the group. There are a few forms in the Silurian and Devonian, but later than this the group dies out. The phyletic picture is not clear. Within a comparatively short period of geological time in the early Ordovician most of the main cystoid groups have appeared, and one must be

Fig. 13   Range of body form in the cystoids

a Diagrammatic reconstruction of a typical cystoid, based on the diploporite *Fungocystis* (see *e*), with cover-plates closed over the food grooves. There is no evidence indicating the length of the brachioles.

b to *f*, evolutionary trends in the Diploporita, showing the progressive lengthening of the food grooves over the theca.
b Aristocystites (Ord.).
c Eucystis (Ord.).
d Glyptosphaerites (Ord.).
e Fungocystis (Ord.).
f Dactylocystis, in which the dipores are restricted to the region of the food grooves.

g to *i*, equivalent trends in the Rhombifera, showing also the restriction of the pore-rhombs to a few areas only.
g Echinosphaerites (Ord.–Sil.).
h Cystoblastus (Ord.) with two pectinirhombs, of which one is shown.
i Staurocystis (Sil.), with three pectinirhombs, of which one is shown.
   The arrows in the diagram do not necessarily imply phyletic lines.

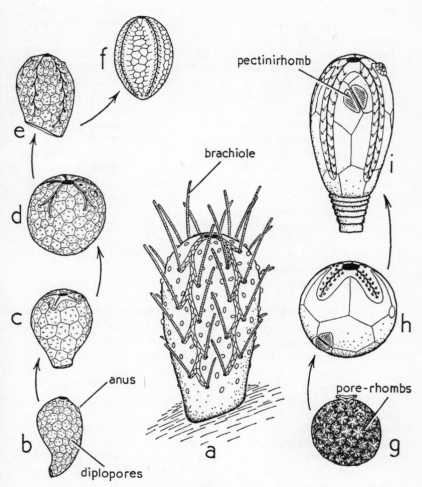

pectinirhomb

brachiole

anus

diplopores

pore-rhombs

Fig. 13

content to trace possible evolutionary lines and morphological trends from the known forms around that time. We have already seen (p. 87) how the very early *Macrocystella* provides a possible foundation for the evolution of the rhombiferan pore structures, and at about the same time there appear the first diploporite cystoids, such as *Sphaeronites* and, slightly later, *Aristocystites* (Fig. 13*b*). In these forms the theca is spherical or flask-shaped and the arm-like brachioles originated from close to the mouth. In others, both diploporites and rhombiferans, the food grooves extend over the theca in various ways, and the grooves are bridged by moveable plates, rather like the lappets lining the food grooves of living crinoids. Thus, in the Diploporita we can trace a series (though not necessarily a phyletic one) starting with forms like *Aristocystites* (Fig. 13*b*); through such forms as *Eucystis* (Fig. 13*c*) in which the brachioles, usually five, arose from only a little way from the mouth; through *Glyptosphaerites* (Fig. 13*d*) in which the grooves are longer and the brachioles are increased in number and arise from branches of the five main grooves; to *Fungocystis* (Fig. 13*e*) where the food grooves extend to the base of the theca. A modification of this last pattern is seen in such forms as *Dactylocystis* (Fig. 13*f*), in which all the dipores are aggregated close to the food grooves, possibly to derive benefit from the currents in that area for increased respiratory efficiency.

In the Rhombifera a similar trend can be traced from forms like *Echinosphaerites* (Fig. 13*g*), in which there are effectively no grooves on the theca, the brachioles arising almost directly from the perioral region; through forms like *Cystoblastus* (Fig. 13*h*), in which the food grooves and associated brachioles are borne only on the adoral surface; to *Staurocystis* (Fig. 13*i*) in which the grooves and brachioles extend nearly to the point of insertion of the stem.

To summarise, the cystoids are two fairly distinct lines of flask-shaped animals, in some of which the theca was attached to the substratum, while in others there was a stem. They held out brachioles to catch the rain of food dropping to the ocean floor. In some the plates were numerous and haphazardly arranged, while in others fewer plates were arranged mainly in whorls of five; in some the brachioles arose from close to the mouth, while in others they arose from the sides of food grooves on the theca; in some the respiratory organs were scattered over

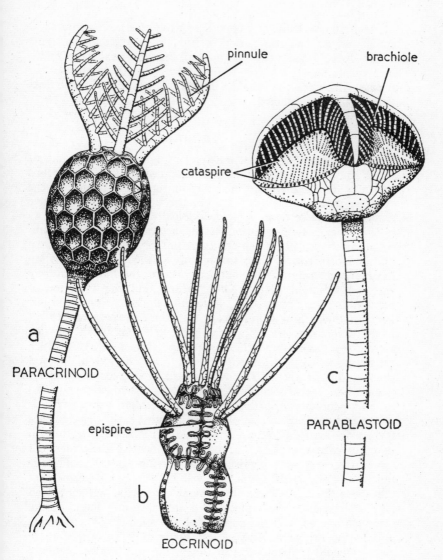

Fig. 14   Lesser-known extinct crinozoans

a A paracrinoid, *Comarocystites*.
b An eocrinoid, *Lichenoides* (redrawn from Ubaghs[73])
c A parablastoid, *Blastoidocrinus*.

the entire theca, while in others these structures were restricted to special areas. The respiratory organs either relied on currents of sea-water flowing through special channels inside the theca, or on coelomic fluid flowing through canals on the outside of the theca; sometimes the canals were of soft tissue and sometimes their walls were calcified.

THE EOCRINOIDEA

This assemblage of rather rare forms has the distinction of being among the first echinoderms to appear in the fossil record (p. 159). Its members are probably more closely related to the cystoids than to the crinoids; in fact, a suggested alternative name for some of them is 'cystocrinoids'.

The theca, often stemless, as in *Lichenoides* (Middle Cambrian), is shaped like an elongated flask (Fig. 14b), with a crown of brachioles surrounding the mouth at the top. There is no flexible tegmen on the oral surface; instead, the theca is entirely enclosed by rigid plates. This feature, and the presence of brachioles rather than arms, distinguishes them from crinoids. Most of them show pentamerous symmetry in the arrangement of the brachioles: the ambulacral pattern is reminiscent of such cystoids as *Eucystis* and *Glyptosphaerites* (Fig. 13c, d), in which the brachioles arise from short branches of the five main food grooves on the theca. The grooves all have protective coverplates. The thecal plates themselves do not always show an orderly, pentamerous arrangement, but are often haphazard.

The feature of the thecal plates which most easily distinguishes the eocrinoids from other crinozoans is the presence of sutural pores or *epispires*. These occur between most of the plates and are interpreted, not surprisingly, as having had a respiratory function. Their structure is not well known, but it seems that they bore 'blisters' of the integument across the wall of which gaseous exchange took place with the coelomic fluid, in the same manner as with the papulae of asteroids (p. 36). In some, apparently, the walls were of soft tissue only; in others, there were embedded plates; and in others the wall was completely calcified, like the fistulipores of some cystoids (p. 84).

The stem often had an expanded distal end, either to act as an anchor or to provide attachment to a hard object; but some eocrinoids lacked a stem altogether, or had at most a mere stump, and these are interpreted as having sat upright in the sediment.

Their phyletic relationships are obscure. It does not seem unreasonable to see in them an affinity with cystoids; indeed, only recently (1968) has *Macrocystella* and its relatives been removed from this class and placed with the rhombiferan cystoids (p. 84). But what of their relations with the crinoids? Some workers see great differences between the plate arrangements of the two groups, and particularly the nature of the appendages (oral brachioles in eocrinoids, true arms arising from radial plates in crinoids). But others see the transformation from one type to the other as fairly straightforward. The fact that eocrinoids first occur before the crinoids may or may not be important (such is the fickleness of the fossil record); the fact that the two groups have such similar basic body form and habit may be coincidental; but the fact that both groups fit so closely the hypothetical archechinoderm derived solely on neontological grounds[137] strongly suggests that they both shared common ancestry not too far removed from the known forms.

THE PARACRINOIDEA

These little-known crinozoans, too, appear to have closest affinity with the cystoids rather than the crinoids, if only because the plates of their wineglass-shaped bodies (Fig. 14a) sometimes have deep folds projecting into the thecal interior, rather like the confluent dichopores of the glyptosphaeritid cystoids (p. 90). There may have been a skeletally-supported integument over the outside of these structures, but so few specimens are known that it is not possible to be certain about the exact nature of the system.

The body does not have a tegmen orally, and therefore they cannot properly be included in the crinoids; yet they have up to four arms or brachioles composed of a single column of plates, and these arms bear pinnules, which is a crinoid-like feature. The thecal plates are haphazardly arranged, and in some forms, such as *Comarocystites*, they have a deep concavity at the centre. The mouth is roofed over by cover-plates, and a single pore has been noticed, which may be a hydropore or gonopore or both. So little is known about them, and they are found over so short a stratigraphical range (Middle Ordovician only) that no valid comment can be made about their relationships, beyond their obvious links with the eocrinoids and cystoids.

THE PARABLASTOIDEA

In rocks of Middle Ordovician age, that is, considerably earlier than true blastoids, there occurs a single genus, *Blastoidocrinus* (Fig. 14c), with thecal pores, the *cataspires*, rather similar to the dichopores of some cystoids and with folds in the plates not unlike those in some blastoids. Though poorly known, this genus does seem to stand apart from true blastoids, and so a separate class, the Parablastoidea, has been suggested to contain it. How justified this is may become clearer as more is known of the animal.

The theca, superficially similar to that of blastoids, has many irregular plates, and the stem is inserted in a deep concavity aborally. The ambulacral structure is unique in that the cover plates have large wing plates arising from them to form a star-shaped raised 'platform' on the oral surface of the theca. Brachioles line the ambulacra in a blastoid-like manner, and in the fossils these tend to be preserved lying against the wing plates, as though forming the sides of the platform. Possibly only the bases of the brachioles are ever preserved, or possibly the brachioles could be moved away from the sides of the ambulacral platform when the animal was feeding. To date, the fossil is too poorly known for its mode of life to be assessed with any confidence; neither can its relationships be suggested with conviction (see p. 159).

THE BLASTOIDEA

This group shows some similarity in general thecal structure and brachiole arrangement to the later cystoids, that is, those that have attained pentamerous symmetry, and some authorities favour placing them as an order of the Cystoidea. On the other hand, differences in the detail of their ambulacral structure and in those structures that are generally considered to have been respiratory in the two groups are so marked that it seems better to regard the blastoids as a distinct class, as is more usual[8].

The theca (Fig. 16a) is bud-shaped, with five food grooves radiating from the upwardly directed mouth; in life, numerous brachioles arose from the sides of its food grooves and sometimes these are preserved *in situ*. The basic plate-arrangement of the theca is simpler than the simplest situation found in the cystoids: three whorls of plates make up almost the whole theca, except for the food grooves. Each whorl consists of five plates: round the mouth is a ring of deltoids, between and below them a ring of radials, usually the largest plates of the theca and notched orally to receive the food grooves, and below these are the basals, which in many blastoids become secondarily reduced by fusion to three plates, two large (each formed by the fusion of two plates) and one small. The stem is attached to the basals and breaks up into roots at its lower end for insertion into the substratum. In the great majority of blastoids the slight irregularity in the basal whorl of the theca and the presence of the anus in one interradius is the only departure from radial symmetry; but in some there is a much greater departure from radiality: one food groove may be reduced (as in *Eleutherocrinus*, Fig. 16g) or modified in structure (as in *Astrocrinus*, Fig. 16h), making the theca bilateral.

The most intriguing feature of the blastoids is the nature of the food grooves and neighbouring structures. Well-preserved specimens, when sectioned transversely in the region of the ambulacra, yield quite a bit of information about the soft parts which once impinged on the skeletal plates. A blastoid ambulacrum, as mentioned before, lies between the deltoid plates orally and in a notch in the oral side of the radial plates aborally. The floor of the ambulacrum is formed by a single elongated plate, the *lancet* (Fig. 15a), and two columns of *side-plates*. Either the side-plates lie beside the lancet, as in *Pentremites* (Fig. 15c) or they come to lie on top of (oral to) the lancet, as in *Orbitremites* (Fig. 15d), so that in this case the lancet can be seen externally only through a tiny groove between the side-plates. In addition, some forms have an extra column, the *outer side-plates*, on each side. In life, and in better-preserved specimens, each of the side-plates bore a single brachiole which contained a longitudinal food groove on its oral side, continuous with the main ambulacral groove. In most blastoids, in common with other crinozoans, cover plates could apparently be closed over the brachiolar and ambulacral grooves, and over the mouth area for protection, but in a few the cover plates are functionally replaced by upstanding spines alongside the ambulacral grooves.

In sectioned specimens, however, one can see further details. In the middle of each lancet plate there is a longitudinal canal (Fig. 15a) which joins a ring-like canal in the oral part of the deltoid circlet of plates. By comparison with living crinozoans, these canals almost certainly represent places where part of the animal's nervous system was contained. Then, alongside each

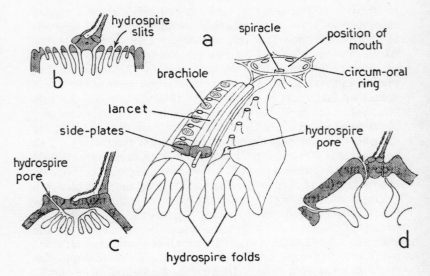

Fig. 15   Structure of blastoid ambulacra

a Schematic perspective diagram of part of one ambulacrum of a typical blastoid, based on that of *Pentremites*. The side-plates of the right side have been removed.

b T.S. ambulacrum of *Codaster*, showing hydrospire slits.

c T.S. ambulacrum of *Pentremites*, showing ingrowth of hydrospire folds. The hydrospire pores alternate, so are shown on the left side only.

d T.S. ambulacrum of *Orbitremites derbiensis*, showing single hydrospire folds. The side-plates overlie the lancet.

In *b*, *c* and *d* a brachiole and the cover-plates to the food grooves are shown on the right side only.

ambulacral groove, one can usually see very thin internal folds of the thecal plates hanging in the central cavity, folds that were clearly calcite-supported in life; they resemble some of the calcitic dichopore canals of cystoids. These are the *hydrospires*, which are almost universally regarded as having been respiratory structures.

The hydrospire system

If one imagines the typical pore-rhombs of a rhombiferan cystoid or the cataspires of a parablastoid (p. 94) brought to lie close to the ambulacral grooves so that they traverse the radio-deltoid sutures in each interambulacral area, one can see how the hydrospire system in the blastoids possibly arose. In the so-called fissiculate blastoids, such as *Codaster*, Silurian (Fig. 15*b*, 16*c*), the situation is closest to that of the cystoids: a series of parallel slits runs at right angles across the radio-deltoid suture, each slit leading inwards to a single fold of tissue supported by a very thin layer of skeletal material (and therefore preserved in the fossils). In spiraculate forms, such as *Pentremites*, Carboniferous (Fig. 15*c*, 16*e*) the folds are pushed into the theca and the connexion to the exterior is no longer by a slit along the length of each hydrospire fold, but is by a single longitudinal row of pores, the *hydrospire pores*, usually placed between the side-plates. In addition, in this arrangement each set of hydrospire folds has a pore, the *spiracle*, near the mouth, though in some two adjacent hydrospire sets share the same spiracle, as shown in Fig. 15*a*. It is most likely that the hydrospire pores provided an entrance to the system for sea-water, the folds a respiratory surface for exchange of gases between sea-water and coelomic fluid, and the spiracles an exit for the water. This interpretation assumes, of course, that respiratory exchange took place through fenestrations in the thin calcite sheet supporting the hydrospire folds; a similar assumption has to be made in interpreting the function of the majority of thecal pore systems in the fossil crinozoans.

While most authorities accept that the main function of the hydrospires was respiration, some see them having additional functions. Perhaps they were similar to the bursae of ophiuroids (p. 54 and Fig. 6), and functioned not only for respiration but also in reproduction, by receiving the fertilised eggs from the gonads and protecting them during the early stages of development. No trace of gonads or gonoducts has been found in blastoids, and the suggestion is that genital products may have been shed into the hydrospires where they would be protected for a time, before being expelled through the spiracles.

The evolution of the blastoids

It is not at all clear how the blastoids arose. They share with several other early crinozoans, such as cystoids and parablastoids, skeletally supported respiratory grooves and canals; yet there are few clues as to their ancestry. When the true blastoid pattern appears, that is, in the late Silurian, both fissiculate and spiraculate forms are found at once, so one cannot invoke time-relations to suggest which group of blastoids is primitive. But it is not un-

D

reasonable to suggest that the fissiculate form of the hydrospire (grooves in the thecal plates) is a less specialised arrangement than the spiraculate form (inhalent pores, internal folds and exhalent spiracle); further, the fissiculate form more closely resembles the respiratory gadgets of other fossil crinozoans.

No definite evolutionary trend can be traced within the group, though the earlier forms do tend to have ambulacral grooves restricted to the upper face, as in *Codaster* (Fig. 16c). Blastoids with grooves restricted to the upper face turn up again and again in later geological periods, but there does seem to be a trend in forms occurring from Silurian to Permian times for the theca to become taller and the ambulacral grooves to extend further and further down its sides until they reached nearly to the stem insertion, as in *Nucleocrinus* (Devonian) and the Carboniferous forms *Orbitremites* and *Pentremites* (Fig. 16e). In the Devonian and Carboniferous, however, there was a slight tendency towards irregularity; the evidence suggests that this happened independently in at least two lines: among the fissiculate forms in the astrocrinids (Fig. 16h) and among the spiraculates in the eleutherocrinids (Fig. 16g). Finally, very late in blastoid history (Permian) some forms had the ambulacral areas on flanges or wings, as in *Pterotoblastus* (Fig. 16f) and the very star-like *Thaumatoblastus*.

So the blastoids as a whole exhibit a remarkable constancy of forms. Summarising the trends, one can trace, first, an elaboration and apparent increase in efficiency of the unique hydrospire system, and, secondly, a lengthening of the food grooves; some forms became slightly irregular by modification of one ambulacrum, while others increased the food catchment area by having the ambulacra thrust out on wing-like extensions of the theca.

Fig. 16 Evolution and range of body form in the blastoids

a Reconstruction of a typical blastoid, such as *Orophocrinus* (Carb.).

b to e, the main evolutionary trends within the blastoids.

b *Codaster* (Sil.), with ambulacra still confined to oral region.

c *Orbitremites*, and d *Pentremites* (Carb.), in which ambulacra extend nearly to stem insertion. Entry to hydrospires by pores.

e *Pterotoblastus* (Perm.), with ambulacra borne on arm-like processes.

f and g, bilaterally symmetrical blastoids, each with one ambulacrum modified.

f *Eleutherocrinus* (Dev.). i) side view, ii) oral view.

g *Astrocrinus* (Carb.). i) side view, ii) oral view.

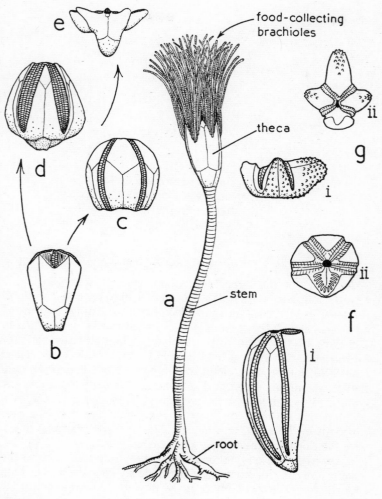

Fig. 16

8

EXTINCT ECHINOZOANS

EDRIOASTEROIDEA
HELICOPLACOIDEA
OPHIOCISTIOIDEA
CYCLOCYSTOIDEA

THE EDRIOASTEROIDEA

This is one of the three echinoderm groups to appear in the lowest Cambrian rocks (p. 118), though it may have had a representative even earlier, in the Precambrian sandstone imprint of an animal called *Tribrachidium*. From what can be deduced from the trace of this once soft-bodied animal, it appears to have looked superficially like a three-armed version of *Agelacrinites* (Fig. 17*h*), but as no skeleton was preserved, we cannot be sure that it was an echinoderm at all. Until this early occurrence is confirmed (which would make *Tribrachidium* the earliest known echinoderm) we must be content to say that the edrioasteroids range from Lower Cambrian to Lower Carboniferous times, with their acme in the Middle Ordovician.

Some early workers[75] thought that the group was close to the ancestor of the Asterozoa and Echinozoa, because it was then thought to be the first group in the phylum to show a tendency towards an eleutherozoic (free living) existence (helicoplacoids being still unknown). But more recently[79] the group has been thought of as a separate line, probably not all that close to the origin of the other free forms.

In spite of the presence of a stem in some edrioasteroids (e.g., Fig. 17e), there appear to be good reasons for including the class among the Echinozoa rather than the Crinozoa. The earliest forms, e.g., *Stromatocystites* (Fig. 17c) lacked a stem, and in addition the ambulacral structure of some of them is reminiscent of that of other early echinozoans, such as helicoplacoids (p. 105). But they do not appear to have had powers of movement. Those with stems were almost certainly fixed in the soft sea-bed, while the discoid forms are usually found attached to the shells, etc., of other animals, as though they may have settled after larval life on the only solid objects available to them.

The upper (oral) surface consists in general of five interambulacral areas separated by the five ambulacral food grooves. The interambulacra consist of a mass of irregular plates which are sometimes imbricate (overlapping, like the tiles on a roof), as in *Agelacrinites*, or abutting (like a tesselated pavement) as in *Edrioaster*, or fused into one large deltoid plate per interambulacrum as in *Cyathocystis*. In some the interambulacral plates show tubercles which most likely bore spines, but none of the spines have ever been found. In one of the interambulacral areas lies the anus, borne on a spire consisting of a ring of plates, sometimes five but usually more.

Close to the mouth, and usually in an interambulacral area, is a third aperture whose function is obscure. Most workers conclude that it is a hydropore, since it is assumed that any plated echinoderm with a well developed water vascular system, which some of these clearly had, must possess an external opening to it. But if this aperture is a hydropore, where is the gonopore? Did the aperture function jointly as a hydrogonopore, or is it indeed a gonopore only, the hydropore having opened internally, as in present-day holothuroids? This question has not yet been resolved.

The food grooves radiate out from a central mouth, which is surrounded by a strengthening frame consisting of five radial and five interradial pieces. In some forms, chiefly those which seem to be primitive, the food grooves are straight, while in others they are curved; the curving may be in the same direction, or, more often, one or two rays may curve in the opposite way, to surround the anus and the mysterious 'third aperture' (Fig. 17g, h). In some forms, e.g. *Thresherodiscus*, the ambulacra branch several times. Both grooves and mouth region were covered by roofing plates, some of which remain in well-preserved specimens; those over the grooves were probably movable, while those over the peristome were fixed.

The food grooves are floored by ambulacral plates. In most groups these are in a double series, but some later forms (e.g. *Lepidodiscus*) have a single series only. Some of the ambulacra, both biserial and uniserial, have pores between adjacent plates reminiscent of those in asteroids (p. 36), and, moreover, in some of the biserial forms a channel runs down the centre of the food groove in the same position as the radial water vascular canal of asteroids. Almost certainly, this arrangement represents the imprint of a tube-foot system, in which the channel once carried a water canal and the pores carried canals to internal ampullae. Unfortunately, some edrioasteroids, such as the agelacrinitids, do not have such pores or channels, even though their preservation would appear to be good enough to show them. Either these forms lacked tube-feet, or the entire tube-foot system was contained in the soft tissues above the ambulacral plates, as in today's crinoids.

One imagines edrioasteroids feeding by opening the cover-

Fig. 17 Structure and range of body form in the edrioasteroids

a Reconstruction of a typical edrioasteroid, with ambulacral cover-plates open and tube-feet extended, based on *Edrioaster*.

b Tentative reconstruction of part of one ambulacrum. The first two cover-plates on each side are open, the rest almost shut. There is no evidence to show how long the tube-feet were. This reconstruction does not apply to the agelacrinitids, which apparently had no ampullae.

c to h, range of body form. All drawn with cover-plates closed.

c *Stromatocystites* (Camb.), probably the simplest edrioasteroid.

d *Cyathocystis* (Ord.), with straight ambulacra confined to oral surface.

e *Astrocystites* (Ord.), a stemmed form, with ambulacra extending further over the theca.

f *Pyrgocystis* (Ord.–Sil.), with sturdy stem enclosed by imbricating plates; ambulacra restricted to oral surface.

g and h, edrioasteroids with curved ambulacra.

g *Edrioaster* (Ord.–Dev.), with one ambulacrum curving in a direction opposite to that of the others.

h *Agelacrinites* (Dev.), with two ambulacra curving the other way.

The arrows do not necessarily imply phyletic lines, though time-relations and comparative structure are consistent with this being a rough family tree of the edrioasteroids.

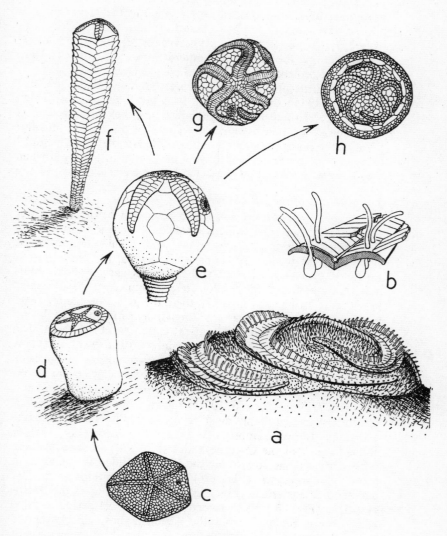

Fig. 17

plates (Fig. 17a, b), extending their tube-feet to increase the catchment area, and possibly even bending the tube-feet in a crinoid-like manner (p. 138) to help food into the groove. When extended, the tube-feet would also serve for respiratory exchange. On retraction, the tube-feet would be brought into the food groove and protected by the closed cover-plates.

Another little puzzle some edrioasteroids pose is the presence of pores between the *cover plates* over the food grooves, which led Bather[75] to suggest that tube-feet may have emerged even when the cover plates were closed over. This seems unlikely: perhaps the pores represent a ventilation system so that a respiratory current could be kept flowing past the tube-feet even when they were withdrawn and protected by the cover plates.

The evolution of the edrioasteroids

The origin of the group must have been well before the Lower Cambrian, so that fossil evidence is lacking. Some authorities[79] see similarities between edrioasteroids and psolid holothuroids, principally on grounds of similarity in general morphology; but no holothuroid has external ambulacra, as early edrioasteroids apparently had, and the food-groove system of edrioasteroids is much more 'crinozoic'. Possibly, then, the group originated from a primitive crinoid, such as a *Stephanocrinus*-like form, by losing arms and pinnules and by a general flattening of the theca. The structure of the crinoid tegmen is certainly consistent with this view, and if those edrioasteroids that show no ambulacral channels or pores (p. 102) really had a crinoid-like water vascular system placed above the ambulacral plates, this adds strength to the argument.

Unless *Tribrachidium* (p. 100) is shown to be an edrioasteroid, the earliest and probably most primitive family is the Stromatocystitidae, Lower Cambrian to Upper Devonian (Fig. 17c). It seems likely that the Cyathocystidae (Fig. 17d) arose from these fairly early on, and so also did the Hemicystitidae, including such forms as *Pyrgocystis* (Fig. 17f), the Agelacrinitidae (Fig. 17h) and the Edrioasteridae (Fig. 17g). A strange group is the Astrocystitidae (Fig. 17e) which passed through a brief period of fame when it was removed from the edrioasteroids, chiefly on the basis of ambulacral pore structure, and placed in a new class of its own, the 'Edrioblastoidea'. Now, however, it is back in this class, since new finds have shown that in most features it closely resembles other edrioasteroids, though the radial water vascular canals were próbably internal to the skeleton. Such migration of the water canals from external to internal happens in another echinozoan group, the echinoids (p. 70).

To summarise, the main trends in the evolution of the edrioasteroids appear to be, first, a tendency in some to prolong the vertical axis so that the food-catching apparatus is elevated well above the sea-floor; and, secondly, a lengthening of the ambulacra, either by curving or by branching. The water vascular canals in some migrated towards the interior of the body.

THE HELICOPLACOIDEA

These unusual cigar-shaped animals (Fig. 18a) also occur in the lowest Cambrian rocks, and are therefore among the first echinoderms. Most of the specimens are assigned to a single genus, *Helicoplacus*. The class stands well apart from the others, its members having spiral pleats on the body and a single spiral ambulacrum making about two turns, with a short subsidiary branch making half a turn close by. The plate arrangement is such that the animal could apparently extend and contract. The plates of the upper (interpreted as oral) part of the body imbricate downwards, while those of the lower (apical) part imbricate upwards. Some of the plates bear spine-like prominences. Unfortunately, none of the main apertures, such as mouth, anus, gonopore and hydropore, have been identified with certainty: even the mouth is only assumed to have been at the upper pole, and wherever it was it must have been very small.

Specimens of another genus, *Waucobella*, have yielded information about the ambulacra (see also p. 145), and it is now fairly certain that there was a tube-foot system arising from the grooves. Durham[76] thinks that the ambulacra may have been solely respiratory, because of their small area in relation to the bulk of the animal; food collection, he suggests, was performed by the interambulacral areas, which passed the particles by ciliary currents to the mouth. The living posture was very likely vertical, with the apical pole inserted into a short burrow in the sea-bed. Then the animal could 'untwist' into its feeding position or spiralise again into the safety of its burrow. In shortening, the movement of the plates relative to one another may have closed the ambulacral grooves to protect the tube-feet. The suggestion of a partial burrowing, upright posture is strengthened by the fact that the ambulacra are confined to the upper part of the body, and the direction of imbrication of the plates changes in the region of the ambulacra, as though the upper part of the body is under different influences from the lower.

The relationships of the helicoplacoids

The early occurrence of this group in the echinoderm fossil record makes it important in any consideration of the phyletic history of the phylum. Yet so unlike any other group is it that its usefulness in making a family tree is reduced. It seems improbable that it left any descendants, but what, if anything, can be said about its origin and relations? There is marked counterclockwise torsion in some edrioasteroids, and slight torsion in the same direction in a few primitive echinoids, such as *Ectinechinus*. But the picture is not clear enough to go beyond mentioning this shared characteristic. Equally tentative is the similarity in plate imbrication between helicoplacoids and psolid holothuroids. What emerges is a faint possibility that helicoplacoids, edrioasteroids, echinoids and holothuroids—the main groups of the Echinozoa—shared common ancestry in a form that had imbricating plates protecting a flexible body. The helicoplacoids appear to have adopted a singularly unsuccessful variation of this body plan.

THE OPHIOCISTIOIDEA

A few dozen curious fossils, assigned to about four genera, have been found in rocks of Ordovician to Devonian age which their

Fig. 18   Lesser-known extinct echinozoans

a A helicoplacoid, *Waucobella* (L. Camb).
b An ophiocistioid. Reconstruction of *Sollasina* (Sil.–Dev.).
c A cyclocystoid, *Cyclocystoides* (Camb.). Two alternative interpretations of this poorly-known fossil.

i, with the tube-feet interpreted as locomotory and possibly feeding organs, directed *downwards*, and the sutural pores of the opposite side bearing respiratory papulae. Mouth in the centre of the under surface.
ii, with the large tube-feet interpreted as respiratory organs and the sutural pores on the opposite side as bearing locomotory tube-feet. The plated apron may have been able to fold over the retracted respiratory tube-feet, as shown on the right side of the diagram. Mouth again in the centre of the under surface. (From an interpretation kindly supplied by Prof. J. Wyatt Durham.)

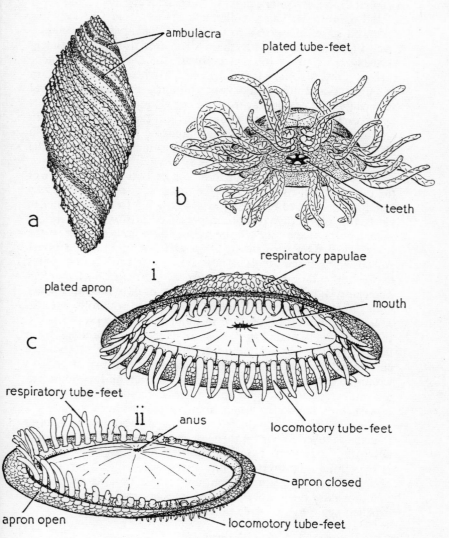

a — ambulacra

b — plated tube-feet, teeth

c — i — plated apron, respiratory papulae, mouth, locomotory tube-feet

ii — respiratory tube-feet, anus, apron open, locomotory tube-feet, apron closed, locomotory tube-feet

Fig. 18

original describer, Sollas, thought to be an order, the Ophiocistia, of the ophiuroids. Later, Sollas and his daughter[81] revised this view and raised the group to Class level. The animals possess a test (Fig. 18b) up to 10 cm in diameter, usually completely encasing the body, with the mouth in the centre of the under surface; on this surface the plates are fairly regularly arranged in ambulacral and interambulacral columns, but on the aboral surface they do not appear to be arranged in any order. The anus opens excentrically on the aboral surface, the madreporite is in one interradius on the oral surface and the gonopore, single in some and multiple in others, is in the same interradius as the madreporite.

The outstanding feature of these fossils is the preservation of the tube-feet, which arose from the under surface only. It is very rare for tube-feet to be preserved at all in any fossil echinoderm, yet in this group they have been preserved *in situ*, and there are even reports of the tube-feet being found without the rest of the body. They were exceedingly large and had imbricating plates embedded in their walls. They arise from alternating pores between the ambulacral plates. In two genera, *Eucladia* and *Euthemon*, there is good evidence that oral tube-feet, smaller than the others and possibly sensory, occurred round the mouth. This is a common occurrence in modern echinoderms (p. 138). In all genera the mouth seems to have been bordered by a flexible peristomial membrane in which many tiny plates were embedded, as in echinoids. The mouth itself had five large jaws within it, radially arranged, which look as though they represent a dental apparatus similar to that in echinoids (p. 64), though how such an apparatus was manipulated is not known.

There seems little doubt that the animals were free-living and that they moved about on their huge tube-feet, the teeth rasping encrusting organisms from the substratum. How they respired is something of a mystery, since there is no sign of respiratory organs. Possibly they underwent anal respiration, as do modern holothuroids, the sea-water being sucked into the cloaca and expelled again by rhythmic pulsations of the peristome.

The relationships of the ophiocistioids

The superficial similarity of these fossils to armless ophiuroids is almost certainly due to convergence. Though the disposition of the aboral plates of some has been said to resemble that of young ophiuroids, the oral plating in the two groups is not comparable. And despite the fact that the madreporite, like that of ophiuroids, is on the oral side, there is a gonopore in one interradius only. There is also a prominent anus, which ophiuroids lack. The similarities with echinoids are much more notable: the spicule-strengthened peristomial membrane and buccal apparatus, the periproct and the single column of interambulacral plates are all echinoid features. But against this there is no apical system and the anus is excentric. So if ophiocistioids and echinoids shared common ancestry, it was far removed from their known members.

THE CYCLOCYSTOIDEA

These discoid echinoderms from the Cambrian are very poorly known; in fact, only one genus, *Cyclocystoides*, is recognised[79]. It consists of a ring of heavy ornamented plates, the submarginals, 'oral' and 'aboral' sheets of thinner plates (the position of the mouth has not been identified with certainty) and a skirt of small plates round the margin (Fig. 18c). The aboral sheet of plates has sutural pores between the plates, and the submarginals have large cup-like pores directed 'downwards', with small canals leading from the base of the canals to the interior of the theca.

Opinions on the organisation of these echinoderms are extremely divergent. There is no agreement on which sheet of plates represents the oral surface, and some authorities[79] regard the sutural pores as having given rise to tube-foot-like extensions of a water vascular system (Fig. 18c, ii), the internal parts of the system being arranged in radiating channels. It must be said that such an arrangement bears no relation to any known water vascular system. More likely, the pores of the submarginal plates alone held external projections of the water vascular system, probably in the form of large tube-feet like those of the ophiocistioids (p. 106), but non-plated (Fig. 18c, i). If so, then the most likely living position would be with the tube-feet directed downwards; it follows from this that the sheet of plates with sutural pores would be the upper surface of the animal, and so most likely the sutural pores gave rise to papula-like structures for respiratory exchange. The animal probably moved across the sea-bed on its large tube-feet (there are facets within the sub-marginal cups which could have been muscle attachments), picking up food in some way from the surface of the substratum.

9

HOMALOZOANS, 'HAPLOZOANS', AND LESSER-KNOWN EXTINCT GROUPS

THE HOMALOZOA

This sub-phylum is composed of asymmetrical echinoderms which used to be included in the former class 'Carpoidea'. There was a period when carpoids became known as Heterostelea, because an arm-like structure in some was misinterpreted as a stem, but with very different structure from other echinoderm stems (the name means 'different stem'). Now, however, this arm has been shown to be a feeding organ, with typical ambulacral structure. There are three classes in the sub-phylum: the Homostelea, formerly the carpoid order Cincta, whose members have a typical stem only; the Homoiostelea, formerly the order Soluta, forms with feeding arm and stem; and the Stylophora, formerly the orders Cornuta and Mitrata, with feeding arm only. There seems little doubt that the three classes are related. Some workers think that the feeding arm and stem are radially homologous, that is, that the two structures have been derived from similar, radially symmetrical appendages, by the loss of other members of the series and a structural divergence of the remaining two, but there seems little evidence of this. The sub-phylum ranges from the Middle Cambrian to the Upper Devonian.

The homalozoan theca (see, for instance, Fig. 19a) generally consists of a ring of rigid marginal plates with an upper and lower sheet of plates, different on the two sides. A large thecal opening opposite the feeding arm (when present) has been interpreted as an anus, surrounded by a flexible periproct. The arm and stem generally arise from the marginal ring. In many forms the thecal plates bear the imprint of internal soft parts, and such markings are often very complex, so much so that one worker[85, 148] has re-interpreted them as representing the imprint of a chordate nervous system, with the feeding arm as the remains of a swimming organ. In this view, therefore, the homalozoans have less characters appropriate to the echinoderms than to the chordates, and a new sub-phylum, the 'Calcichordata', has been proposed for them.

This interpretation has not found acceptance. In the first place, there is nothing to indicate that the thecal markings all contained nerves; in the second, it is difficult to see why an animal so similar in grade of organisation to other thecate echinoderms should require a nervous system so markedly more complex than others; and in the third, there is a perfectly good interpretation possible of the structure of the arm (='tail') in conventional echinoderm terms. This is not the first time these animals have been thought of in connexion with the origin of chordates. Gislen[147], too, considered that certain thecal pores were the first appearance of gill slits in the chordate line (see also p. 169).

THE HOMOSTELEA

In some ways these homalozoans, including such forms as *Gyrocystis* and *Trochocystites*, are the simplest of the group, but they are by no means easy of interpretation. There is no feeding arm: the only projecting structure (Fig. 19a) is clearly a stem or peduncle, slightly flattened on the side that probably lay on the sea-bed. The plates of the upper integument have sutural pores between them, possibly once the site of papula-like structures (p. 36). The thecal margin has two openings in it: a large one at the opposite end to the stem and a smaller one to one side of the larger. The larger one has a hinged opercular plate on the upper side of it which looks as if it could have been opened by muscles, and the smaller one has either one or two grooves, guarded by cover plates, leading to it round the margin of the theca. There have been several different interpretations of these openings, particularly before the true nature of the thecal grooves was reported. It now seems most likely[86] that the larger of the two openings was the anus, and that anal respiratory currents could pass in and out of the orifice when the operculum was open. The smaller was probably the mouth, and the grooves leading to it may have had ambulacral structures within them. But in some forms the grooves are so short in relation to the size of the body that there may have been extensions of the ambulacral structures beyond the limits of the grooves. Alternatively, the whole upper surface may have been used for food collection, the food particles being carried over the side of the theca by surface ciliation to the grooves, thence to the mouth. This arrangement may have been necessary to 'steer' the food past the anus while the respiratory currents were being created, and to make sure that faecal matter did not join the procession to the mouth.

THE HOMOIOSTELEA

This group contains forms like *Dendrocystoides*, which have a stem and a plated feeding arm (Fig. 19*b*). The stem widens out where it joins the theca, and the hollow cavity enclosed by its plates may have contained a structure similar to the crinoid 'chambered organ' (p. 30), which is mainly a nerve centre. The anus lay to one side of the stem origin. The arm has a double series of brachial plates and an upper double series of cover plates which could probably be opened, at least distally. One assumes that ambulacral soft parts underlay the cover plates in the usual echinoderm manner. The mouth has not been identified, and probably opened adjacent to the origin of the arm, and may have been roofed over by plates continuous with the series of

Fig. 19   Homalozoans, 'Haplozoans' and lesser-known extinct groups

a to *d*, Homalozoa
a Trochocystites (M. Camb.), a homostelean; the only appendage is a stem.
b Dendrocystoides (M. Ord.), a homoiostelean, with stem and feeding arm.
c and *d Cothurnocystis* (Ord.), a stylophoran, with feeding aulacophore.
    c, underside; *d*, upper side, showing row of special pores, possibly gills.
e Camptostroma (L. Camb.), a poorly-known echinoderm.
f Lepidocystis (L. Camb.), also poorly known.

g and *h*, 'Haplozoa'.
g Cymbionites (M. Camb.), a cycloid, in ventro-lateral view.
h Peridionites (M. Camb.), a cyamoid, in ventro-lateral view.

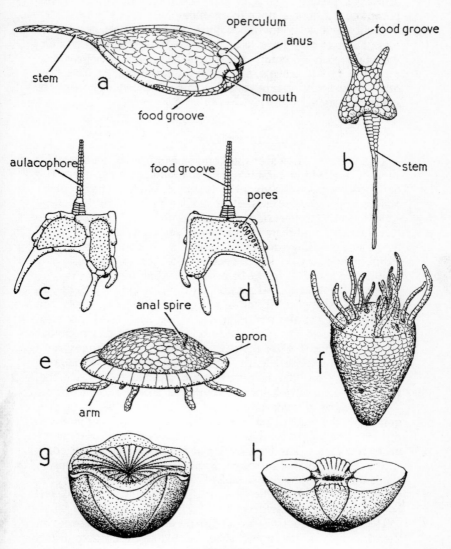

Fig. 19

arm cover plates. Small apertures, which may have been hydropores and gonopores, open to one side of the arm.

As in homosteles, the theca appears to have been flexible, so
these animals may have undergone anal respiration. Certainly,
the anal region has complex plating which might have been concerned in such a process, and there do not appear to have been
other structures, except possible tube-feet, which might be
interpreted as respiratory.

<h2 style="text-align:center">THE STYLOPHORA</h2>

This is the best known of the homalozoan classes, and contains
some of the oddest echinoderms yet found. In most, there is a
flattened theca consisting of a marginal frame of large plates
and an upper and lower sheet of smaller ones, but in the earliest
stylophoran, *Ceratocystis*, the thecal plates all contribute to
thecal rigidity, and there is no frame as such. The stylophorans all
have a single feeding arm, but on some of the bizarre forms there are
struts, buttresses and studs projecting from the frame plates, mostly
on the opposite side to the arm, and pointing the other way.

One of the most typical members is *Cothurnocystis* (Fig. 19*c*, *d*),
which has the typical frame of marginals supporting upper and
lower sheets. On the underside the frame has a diagonal strut
from one side to the other, and on the same side the frame has
several prominences like the studs on a tea-tray which must have
dug into the sea-bed slightly. Extra anchorage was provided by
projections from the marginal frame parallel to the sea-bed,
which may also have helped to counterbalance the feeding arm.

As in other homalozoans, there is no orifice which is une-
quivocally the mouth, but it is assumed to have been adjacent to
the insertion of the feeding arm and to have been covered by
protective plates. The large orifice at the other side of the body,
with a flexible, plated surround, was almost certainly the anus. A
single thecal opening to one side of the origin of the feeding arm,
either on the upper or the lower surface according to genus, has
been thought of as a hydropore.

The most unusual feature about these animals is a series of
thecal pores on the upper sheet of plates of cornute forms
(Fig. 19*d*). In the earliest cornute, *Ceratocystis*, these are in the
form of simple sutural pores piercing the theca; in the cothurno-
cystids they are elliptical *cothurnopores* each bounded by two

special 'U'-shaped plates and apparently covered by a flexible valve of spicule-supported tissue; in the scotiaecystids, they are closely-set, slit-like *lamellipores* together somewhat resembling a cystoid pore-rhomb (p. 86). These pores have been assigned many functions by various workers, and the puzzle is still largely unsolved. The most likely function appears to have been respiration, and the fact that they are located in one region of the theca only lends strength to the suggestion[84, 85, 86] that they are pharyngeal gill slits, opening internally into the first part of the alimentary tract and providing an exit for water taken in with the food *via* the arm. The effluent water would have passed over respiratory structures before passing out: indeed, the inner rims of some lamellipores appear to have had soft tissue attached to them. The valve-like flap over the cothurnopores would have been lifted to allow the effluent to escape by the slight positive pressure within the animal's pharynx.

There are no cothurnopores, lamellipores or sutural pores of the cornute type in the other order of stylophorans, the Mitrata, but there are simple pores which pierce the thecal plates to open internally at grooves in the inner face of the plates, or sometimes opening into canals in the substance of the plates. Since the internal channels and canals leading to these pores also originate at the anterior end of the theca, they too have been thought of as exits for effluent water, functionally replacing cothurnopores, etc., of the cornutes. But there is neither proof of this nor anything to suggest alternative function.

The feeding arm, or *aulacophore* ('groove-bearer'), has a structure consistent with its having borne a typical echinoderm ambulacrum. Along its length it bears grooves and cavities which suggest that it carried a tube-foot system, protected by cover plates. The aulacophore is regionally differentiated, and it is possible that only in the distal parts could the cover plates open. The whole structure appears to have been capable of bending in all planes.

What strange fossils these are! There is no trace of symmetry, and they show structures found in no other echinoderm group. No wonder so many interpretations of their nature have appeared.

THE 'HAPLOZOA'

Two little-known groups, the Cycloidea and Cyamoidea, were originally proposed as classes within a new sub-phylum of the echinoderms, the Haplozoa, by their author, Whitehouse[87]. Each class was erected on the strength of one fossil genus, the Cycloidea containing *Cymbionites* (Fig. 19*g*) and the Cyamoidea *Peridionites* (Fig. 19*h*). They both come from the lower part (*Xystridura* Zone) of the Middle Cambrian of Queensland, *Cymbionites* ranging through forty feet and appearing in colossal numbers in some bands, while *Peridionites*, twenty-four feet above, ranges through five feet and is far less numerous. They have been included in the phylum solely on the nature of the skeleton: their plates are apparently made up of single crystals of calcite and show the tell-tale reticular microstructure so characteristic of echinoderms (p. 123). But they are so different from previously known echinoderms that they cannot easily be included in any known class.

Cymbionites consists of a dome-shaped, radially symmetrical theca about 10 mm long composed of a radial series of wedge-shaped plates, usually five in number; the concave under-surface is ridged and grooved in a manner which suggested to White-house that the soft parts of the body, with straight gut and coelomic pouches, etc., were all contained beneath the theca, held in place by radially-running ligaments inserted into the grooves which can be seen on the concave surface of the fossils. *Peridionites* also has a dome-shaped theca of five plates, but they are not arranged pentamerally and in consequence the symmetry is biradial. The under surface is again concave, but here White-house saw the possibility of it having housed a bilateral animal, possibly something close to a hypothetical ancestor of the whole phylum, the *dipleurula* (p. 167), which was fashionable in Whitehouse's day.

The relationships of the 'Haplozoa'

It need hardly be said that on such slim evidence as the structure of their very peculiar tests *Cymbionites* and *Peridionites* have proved difficult to slot into the accepted classification of the phylum. At present, it is not really possible to make any reasoned reconstruction of their soft parts or to relate them to any known groups. Some authorities think they represent the reduced thecae of eocrinoids (p. 92) or cystoids (p. 83), or that they are not echinoderms at all. If they are not, they represent another phylum which has adopted the reticular microstructure of a calcite skeleton (p. 123). Until more information is available, all is speculation.

THE CAMPTOSTROMATOIDEA AND LEPIDOCYSTOIDEA

Both these little-known groups occur together in the *Olenellus* Zone of the Lower Cambrian of Pennsylvania. Both have a plated body with flexible plated arms in one or more circlets round the mouth[86]. *Camptostroma* (Fig. 19e) was originally said to be a medusoid coelenterate, because the specimens found first were mere imprints; but then, by good fortune, a specimen came to light with the supporting plates preserved, and the reticular microstructure of the calcite betrayed their true affinity with the echinoderms. In this form, the body is umbrella-shaped, with a ridged apron round its periphery. The arms arise from the periphery, though their number is uncertain. The mouth is on the underside, the anus on the upper, and many of the body plates have sutural pores between them, particularly on the oral surface.

In *Lepidocystis* (Fig. 19f), the mouth and anus are interpreted[86] as having been on the same side, a slightly domed, tegmen-like surface with pores between the plates, as in *Camptostroma*. The aboral surface appears to have been conical in life and covered by imbricating plates devoid of sutural pores. The arms, arising from the oral surface, are in three or so whorls of five arms each. The arm structure of *Lepidocystis* is better known than that of *Camptostroma*; it consists of two columns of heavy plates aborally and two of cover plates orally—a typical echinoderm arrangement.

It seems possible that both forms had papula-like 'blisters' (p. 36) emerging through the sutural pores for respiration, and that a typical ambulacrum, with water vascular canal and tube-feet, occurred on the oral side of each arm, protected by the cover plates. *Camptostroma* most likely lived like a sessile medusa such as *Cassiopeia*; its plating would make it too cumbersome to be pelagic. It might have rested on its oral surface on the sea-bed and picked up detritus with the tube-feet on its arms, or it might have inverted so that its mouth was directed upwards. *Lepidocystis* also probably lived with its mouth upwards, with its aboral point thrust into the sea-bed; in other words, both these forms may have retained the upward-directed posture which is here regarded as ancestral (p. 159).

PENTAMEROUS SYMMETRY AND THE

ECHINODERM SKELETON

The origin of pentamerous symmetry in the echinoderms

One of the main characters separating the echinoderms from other phyla is the presence in all but a few extinct groups of five-fold symmetry. In present-day forms, though some, such as holothuroids and irregular echinoids, have bilaterality superimposed, the basic pentameric body pattern is clearly visible in all classes. Why this relatively constant feature should have arisen has never been adequately explained. D'Arcy Thomson, in his classic work *On Growth and Form*, quotes the mathematics of the arrangement, but makes no attempt to explain its origin; Breder[88] also treats the shape geometrically only, though he does suggest that the significance of pentamery may lie in the ease with which radial growth can occur from a regular pentagon to give a star-shape.

Echinoderms have the sort of skeleton that fossilises easily, and one would expect some help from the fossil record in determining how and when pentamerism arose. The picture is rather muddled, but it does seem that pentamerism was not nearly so firmly established in early groups as it is now. Three echinoderm classes appear together at the beginning of the fossil record in the Lower Cambrian: eocrinoids, edrioasteroids and helicoplacoids. The first two of these were pentamerous then, but the helicoplacoids showed no sign of radial symmetry. It is possible that one of these classes, the edrioasteroids, occurred even earlier than the Lower Cambrian: it may be represented by the strange Precambrian imprints of an animal that has been called *Tribrach-*

idium, from the Ediacara sandstones of Australia. This animal looks very much like an edrioasteroid, but it apparently had only *three* arms. Such an occurrence tends to support Bather's[1] idea that the original echinoderm was tetramerous, but lacked one ray where the anus opened. Of the remaining three, said Bather, the two lateral ones each divided to give the five of typical echinoderms. So *Tribrachidium*, if it is an echinoderm at all, would represent the intermediate three-rayed condition.

The Homalozoa (p. 110) first appear in the Middle Cambrian, and are asymmetrical throughout their range. It has even been suggested that they are not echinoderms (p. 111), though this seems unlikely, possessing as they do the typical echinoderm skeletal structure and apparently a typical echinoderm feeding arm. Later in the time-scale the Cystoidea (Ordovician to Silurian) appear, and in this group we can trace what appears to be a second attainment of pentamery in the thecal plates. At first they show the typical pelmatozoan condition of radially symmetrical theca with the mouth in the centre of the upwardly directed surface and the anus on the side of the theca but with a haphazard arrangement of thecal plates (e.g. *Aristocystites*, Fig. 13*b*, and *Echinosphaerites*, Fig. 13*g*). Later, the theca tends to have fewer plates arranged mostly in cycles of five (e.g. *Staurocystis* Fig. 13*i*). The ambulacra, too, show a tendency to assume pentamerism: there were apparently only two food-collecting brachioles in the early cystoids such as *Pleurocystites*, three in *Allocystites* and four in *Echinosphaerites*, in which the brachioles arose from close to the rim of the mouth, but in those in which the food grooves extended over the theca and the brachioles arose from facets along the sides of the food grooves and further from the mouth (e.g. *Glyptosphaerites*) the number of grooves appears to have settled at five. But there is one very curious thing about the thecae of those animals, such as *Aristocystites* and *Echinosphaerites*, in which the main body plates were irregular and non-pentamerous: the anal pyramid was very often made up of five small plates, and this was the only part of the animal which showed any trace of pentamerism, so far as we can see.

The development of a calcareous theca is obviously to confer protection on the settled echinoderm which cannot move at all, or only very slowly, to escape predators. But before the theca has become strong and efficient as a protection, that is, just after metamorphosis when the plates are not properly bound together, a nudge from a passing animal may be sufficient to dislodge them. It seems to be the case that the chief planes of weakness in any developing theca lie along the inter-plate sutures, so it is desirable to have as few sutures as possible, and these as short as possible, and it appears likely, from all the evidence available, that these two mechanical demands may have influenced the occurrence of pentamerism in echinoderms. That plates do become dislodged occasionally during early development is obvious from the recent and fossil monstrosities which turn up from time to time: in most cases it seems to be the sutures which have given way. In these cases, only if the damage is fairly slight and does not completely upset the subsequent plate growth can the animal survive to maturity.

The tests of all living classes of thecate echinoderms have a basically similar mode of development in their apical (aboral) regions. A central plate, usually with the anus opening through it, is laid down and surrounding it is a ring of five plates, the basals of crinoids and the genitals of asteroids, echinoids and ophiuroids. Gordon[89] has shown that during the development of the sea-urchins *Psammechinus miliaris* and *Echinarachnius parma* the first six plates to form during metamorphosis (the anal plate and the ring of five genitals surrounding it) occupy nearly the whole of the aboral surface, and hence this shield will be required to take a large part in protecting the young urchin from the rigours of the sea-bottom. After the first ring of plates, which are interradial in position, a second ring, the radials, is laid down outside and alternates with them; from this primordium, the other plates of the theca are budded off in various ways, and it follows that if the theca is to be radially symmetrical each ring will have to be composed of plates of identical shape and size. Assuming that the plates must cover the theca completely with no gaping holes between them, one can put forward a strong case for the use of five plates in the first ring. Suppose the number of plates in the first ring were three, then each plate would be roughly of the shape shown in Fig. 20a, and there would be three main lines of weakness, such as the one shown by the arrows. The angles between the continuations of these lines (ideally 150°) are so obtuse that for practical purposes the lines may be considered nearly straight. In other words, though the sutures are few, they are long. If the first ring consisted of four plates (Fig. 20b), there would be two main lines of weakness across the apical

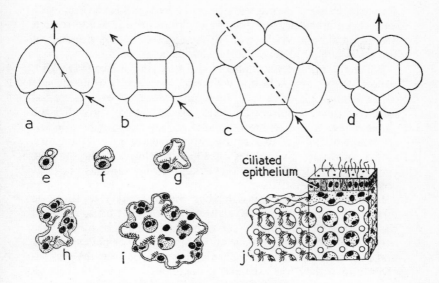

Fig. 20   The echinoderm skeleton

a to *d*, diagrams to show a suggested reason for the establishment of
   five plates in the first ring of the apical disk of an echinoderm.
a With three plates, and a line of weakness shown by the arrows.
b With four plates, opposing sutures in line.
c With five plates, the actual situation, in which no suture is in line
   with any other.
d With six plates, opposing sutures in line.

e to *j*, the formation and structure of the echinoderm skeleton.
e An ossicle-producing cell laying down the first grain of calcite as a
   single crystal in an extracellular matrix.
f The crystal enlarges.
g The cell divides and the crystal grows, but the identity of the matrix
   is being lost.
h Further division and growth; the cell-mass is a syncytium.
i Typical reticular pattern now recognisable.
j To show the relation of the skeleton, the tissue which forms it and
   the external epithelium.

region, one of which is indicated in the figure; although both these
lines are interrupted by the central plate, the two interplate sutures
of each are in a straight line, and if the anus opens through the
central plate this will further add to the weakness of each line.

Similarly, if the first ring were composed of six plates (Fig. 20*d*), each perradial suture and the suture opposite to it would form a line of weakness. Also, with this arrangement the number of sutures is high, and this, clearly, should be avoided.

But if the number of plates in the first ring is five (Fig. 20*c*), none of the sutures is in line with any other, even on the far side of the central plate; in fact, the continuation of each perradial suture will exactly bisect the plate opposite to it, provided the plates in the first ring are of uniform size, and nowhere in the arrangement do any two sutures abut linearly or anything like it. So only with five plates in the first ring are the sutures, and hence the lines of weakness, kept as few and as short as possible[90].

In most echinoids and asteroids the first plate to form on metamorphosis is that through which the anus opens, followed by the ring of five plates surrounding it. This being so, one wonders if it is not possible that the anal group of plates was the first to form in the metamorphosing cystoid, and for the reasons given above they too assumed a pentamerous plan. Of course, few inferences can be made about the way in which the cystoid theca developed after metamorphosis, but it is difficult to see another reason why the anal spire should be pentamerous in the early forms where nothing else is.

One would suppose that the use of pentamerism in preventing each of five points from interfering with the others on the opposite side of an enclosed region would be shown by other groups of the animal kingdom, and indeed this does seem to be so in the Priapulida, the only other phylum exhibiting any marked pentamerism. In *Priapulus caudatus*, for instance, the buccal teeth are arranged in concentric circles of five, and this very curious arrangement has never been adequately explained. On the argument put forward above, it seems likely that here also it would be a disadvantage for the teeth to be in rings of two, four or six, for with any of these arrangements each tooth would interfere with the one opposite when the animal retracts the pharynx to force in the prey. With three or five teeth in each ring, however, this interference would not occur, and only with five teeth in each ring would the aperture be anything like circular. It is particularly interesting that in those polychaetes which also use a toothed pharynx to feed in their prey, such as *Nereis*, the pharyngeal teeth are also in five main groups.

The echinoderm calcite skeleton

In addition to possessing an almost unique type of symmetry, the echinoderms have another feature in which they stand alone in the animal kingdom: the nature of their calcite skeleton. What is its curiously unique structure? Why should it be restricted to one phylum? And could it have anything to do with the adoption of five-rayed symmetry?

The answer to the first question is this: each skeletal element, be it a plate of calcite helping to make up the theca, a spine or one valve of a pedicellaria (p. 130), is apparently composed of a single crystal of calcite (but see p. 125). In the echinoderm larva, and almost certainly in the adult as well, each crystal starts as a tiny grain in an extra-cellular matrix or template adjacent to a skeleton-forming cell or cells (Fig. 20d). In the larva, where needle-like spicules are required, the template elongates as the skeleton-forming cells multiply, and the spicule grows into a needle within it, following the path laid down by the template[92]. But the primordium of the adult skeleton usually accresces in several directions, accompanied by multiplication of the skeleton-forming cells (Fig. 20g, h, i), and although it does not appear to have been observed, one assumes that the accretion takes place in much the same way as has been reported in the larva. In this way a crystal network is formed, and the nuclei of the skeleton-forming cell mass, which appears to be syncytial (without cell walls dividing one cell from another), lie in the spaces within and around it (Fig. 20j). So the crystal can be added to in any direction and can ultimately assume any shape that is needed. The reverse process can happen too: spicules can be eroded away in places if their shape needs to be altered during the growth of the animal. As an example of this, the plates round the peristome of most echinoids are altered in shape by resorption as the animal grows[89].

The reticular pattern possessed by the skeleton has three possible advantages: first, it will be relatively lighter and more economical in material than a solid crystal of the same dimension; secondly, the holes in it will provide an insertion for the connective tissue which binds the elements together; and, thirdly, the holes may offer a resistance to a shearing or splitting force by providing an interruption in the cleavage plane of the crystal. If you snap a plate or spine from a recent echinoderm in two, it will seldom break across a single cleavage plane but will rather fracture along an irregular path, depending on the different thicknesses of the component struts; and the broken ends of the struts will show the same cleavage *angle* but at different *planes*. The situation is rather different, however, where a fossil spicule is concerned. Here the original calcite is replaced first, and then the spaces within the crystal, where the skeleton-forming cell mass originally lay, become filled with secondarily deposited calcite; this attaches itself to the calcite already there along the existing crystal plane, so that now the skeletal element becomes a more or less solid continuous crystal which may even be continued beyond the limits of the original element. In consequence, one would expect a fossilised echinoderm plate to fracture along a single cleavage plane, unlike a plate from a Recent form, and this is indeed the case. In addition, in a well-preserved fossil the original reticular pattern will still be visible, so that palaeontologists are justifiably confident of their ability to put echinoderm remains into their phylum with some certainty.

The answer to the second question, why only the echinoderms have this type of skeleton, does not seem to have been answered. The fact is that other animal groups which utilise calcite as a skeletal material either use solid spicules, as in the sponges, or incorporate extraneous material, such as sand, clay and other chemicals, into the skeleton, as in protozoa, or hold the crystals together with organic fibres, as in molluscs.

The further question may be asked: is there any evidence that other groups using this skeletal system have been evolved independently in the past? And in answer one could point to a little-understood group called the Machaeridia[83], which possesses one genus, *Lepidocoleus*, in which the calcite plates show the single cleavage plane and reticular pattern of the original skeletal material. Indeed, on the basis of the plate structure of this one genus the machaerids have been included in the echinoderms until recently, and the first three editions of this book contained a section on them, in tentative and highly speculative terms. It now appears unreasonable to include them among the echinoderms at all, so different are they in general body structure. Until more is known about them they must remain of uncertain affinity.

It is also possible that the so-called 'Haplozoa' (p. 115) represent another phylum with this special structure. The body plan does not resemble other echinoderm groups, at least so far as interpretations from the skeleton are concerned. But in this group, unlike the machaerids, all known members share the echinoderm-like reticular structure of the skeleton.

The answer to the third question, whether there is a possible connexion between the unique spicule structure and the unique five-rayed symmetry, is probably yes. The strength of the thecal plates composed of single crystals in a reticular shape is most likely such that the weakest places, particularly on the developing theca, are the sutures between the plates; and it has already been shown in the previous section that this is one of the prerequisites of the most plausible explanations for pentamery.

There has been recent controversy on whether the calcite skeleton of which each echinoderm element is composed is really a single crystal, or whether it is built of many microcrystals, shaped like needles, held together in accurate alignment either by an organic matrix or merely by close juxtaposition[91]. Crystallographic studies suggest a single crystal but do not rule out a polycrystalline state: experiments involving removal of the calcite with acid and examining the remains are inconclusive; studies with the electron microscope, both high-resolution transmission instruments viewing a surface replica and lower resolution scanning instruments viewing directly, have been variously claimed to support both views. The matter is still in dispute and must await conclusive demonstration one way or the other.

SPINES AND PEDICELLARIAE

Spines

The echinoderm structures we shall consider in this chapter are those to which the phylum name refers (Greek 'spiny skinned'). Of the living forms the asteroids, ophiuroids and echinoids possess spines, while we can be sure that their predecessors, the fossil edrioasteroids, also possessed them, because the tubercles which bore them are preserved on some specimens. It is not at all certain that the earliest echinoderms had spines: tubercles have not been found in the extinct crinozoans and homalozoans, and there are no spines on crinoids of today, and no evidence that they ever had any. The holothuroids may have secondarily lost them, if we regard them as having evolved from the same ancestral group as the other echinozoans.

The simplest spines are found in the asteroids. Most asteroids have spines on their adambulacral plates, many have them on the other lateral and also the aboral plates, but none has spines in the ambulacral areas. The spines (Fig. 21*b*) consist simply of a calcareous piece held in an erect position on the underlying plate by a number of muscles and covered by the general body epithelium, though this may get secondarily worn off distally. Gland cells secreting mainly mucus but sometimes also poison may be embedded in special spaces in the calcite material. A second type of spine is found chiefly in the Phanerozonia. In Chapter 3 it was mentioned that important currents pass from the aboral surface of most burrowing asteroids to the oral surface both for respiration and for feeding. This current is largely created by special ciliated spines, *clavulae*, in the interstices of the marginal plates on the arm-sides. Each clavula (Fig. 21*c*) is somewhat club-shaped, may have mucous glands distally and has two bands of cilia down its length, one on either side, so that currents are created in one direction only. Also in the phanerozone starfish are special groups of spines on the aboral surface, part of structures called *paxillae*. These structures consist of special raised body ossicles, each of which has a crown of spines arranged in such a way (Fig. 21*d*) that they can be raised vertically or lowered through a right angle to form a covering to the aboral surface. Gland cells in the spines exude a mucus which form a sheet across the closed paxillae so that a cavity can be kept open around the burrowing animal for respiration, etc. Another feature is found in some phanerozones: some spines in forms such as *Pteraster* are held to their neighbours by a membrane (Fig. 21*e*) which presumably can be moved like a fan to create currents for feeding and respiration; then in *Hymenaster* the marginal spines, with fans, form membranes between the arms, possibly to increase the surface area for food collection. In the asteroids it can be difficult to tell the difference between spines and pedicellariae, but these latter structures will be dealt with later.

The spines of ophiuroids (Fig. 21*i* to *l*) are almost entirely restricted to the lateral plates of the arms. They are very like the spines of asteroids in structure (we are coming to expect this of most ophiuroid features!), with the exception that here, unlike their forbears, the spines are not all straight and fairly smooth but may be hooked or thorny, as in *Ophiothrix* (Fig. 21*l*) and *Gorgonocephalus*, presumably to help the animal to get purchase on the substratum or in anchorage; or they may be shaped like an umbrella, as in *Ophiotholia* (Fig. 21*j*), possibly for use on softer substrata.

It is among the echinoids that we find the most highly developed spines. In this group, at least in certain members, they are put to a greater variety of uses than in other groups; they may be used for locomotion, digging, protection, burrow-building, production of currents, breaking the force of waves, bearing poison glands and harbouring the developing larvae. In at least one urchin, *Diadema*, the spines show metachronal rhythm during locomotion, and they move the animal across the ocean floor with considerable speed. The spines of echinoids may be large (*primary* spines), medium (*secondary*) or small (*miliary* spines), though there is no clear division. All are borne on special tubercles on the thecal plates (Fig. 21*a*), each tubercle consisting of a central *mamelon*, on which the concave proximal end of the spine articulates like a ball-and-socket joint, and a saucer-like *areole*, to which the tissues operating the spine are attached. Just below the epithelium is the spine-moving muscle, and below it is a tissue which is predominently collagen, with perhaps some contractile cells interspersed. This so-called 'catch apparatus'[99] functions to clamp the spine under certain conditions of stimulation, apparently by viscosity changes between the collagen fibres. Round the base of each spine is a nerve ring, a swelling of the general basi-epithelial plexus surrounding the body.

It is beyond the scope of this book to describe the enormous range of echinoid spines, but the following is a selection of the more interesting ones. Many echinoids, particularly those inhabiting coral reefs in tropical waters, have large spaces in the calcite material of the spine, in which they house poison. Sometimes the spines are needle-sharp, and easily enter the skin, where they break off to release the poison—and very painful it is too. One subtropical echinoid, *Colobocentrotus*, has special flat-topped spines on its aboral surface (Fig. 21*f*) forming a sort of false test to the body, apparently for protection against wave-action; it seems likely that the Triassic *Anaulocidaris* had a

Fig. 21   Echinoderm spines

a Diagrammatic longitudinal section of an echinoid primary spine and the tubercle which bears it.

b to e, asteroid spines.
b Simple spine borne on the surface of an ossicle, with no tubercle.
c Side and plan view of a clavula, showing bands of cilia.
d Section of four paxillae from aboral surface of a phanerozone. Two on the left are open, two on the right closed. Four papulae are included, out of the plane of the section.
e Fan-spines of *Pteraster*.

f to h, echinoid spines. See also a.
f Flat-topped protective spine of *Colobocentrotus*.
g Paddle-shaped spine of a spatangoid.
h Club-shaped spine, possibly poisonous, of a fossil cidarid.

i to l, ophiuroid spines.
i Simple lateral spines.
j Thorny spines, and k, umbrella spines of *Ophiotholia*.
l Hooked spine of *Ophiothrix*.

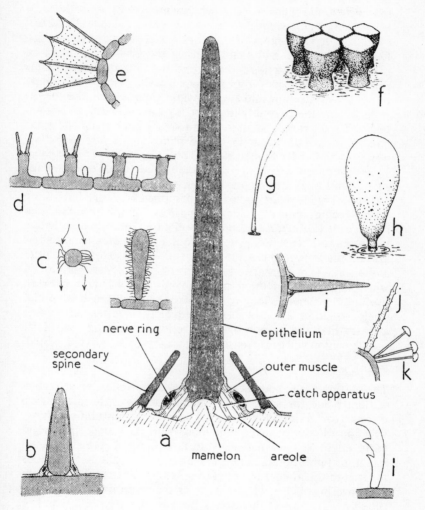

secondary spine

nerve ring

epithelium

outer muscle

catch apparatus

mamelon   areole

Fig. 21

similar adaptation[53]. The genus *Cidaris* and urchins close to it
have relatively huge spines, mainly used for locomotion. But in a
few of them the ends are expanded in a number of ways, the most
usual being a club-shaped form bearing poison glands (Fig. 21*h*);
a cidaroid having spines of this sort was common in the British
Upper Cretaceous, and some levels of the Chalk bear many of
these spines, looking like Indian clubs. In the spatangoids[60] some
spines are oar-shaped, for pressing against the particles of a sandy
or muddy substratum, and also for wiping mucus on to the burrow
walls to stop them from falling in. Clavulae also occur in spatangoids. These, closely similar to the spines here given the same
name in asteroids, occur in special bands called *fascioles*, diag-
nostic of the order Spatangoida. These aggregations of ciliated
spines produce currents in local parts of the animal's burrow for
such purposes as feeding, respiration and excretion.

The last spine adaptation to be described is a very interesting
one: all echinoids except cidaroids possess tiny spherical modified
spines called *sphaeridia*, each borne on a tiny tubercle usually in a
pit in the test. They are generally found in the ambulacral areas
around the mouth, but some urchins have them along the entire
ambulacra. They appear to be organs of balance, since an urchin
which is deprived of them and then inverted will take between
five and ten times as long to right itself as an untouched urchin.
Their structure has not been adequately investigated, but it
appears that the weight of a sphaeridium hanging out of its
usual position will cause sensitive areas at its base to be stimulated.

The pedicellariae

The outside of the body of many echinoderms offers an attractive
settling place for some larvae. Echinoderms with soft bodies,
through which ripples of muscular activity are constantly passing,
will not allow settlement, but those with rigid bodies, or rigid
parts, might well provide a hard surface on which a sessile animal
might metamorphose, and such a visitor would be a serious
menace to the animal by interfering with the external epithelium
and covering such vital soft parts as the respiratory organs and
tube-feet. To protect themselves from such a possibility, and to
help rid the body of any particle that might settle on it, the two
living classes with non-flexible exterior surfaces, the echinoids
and asteroids, are provided with almost unique pincer-like organs,
the *pedicellariae*, among the spines on their bodies. Even such

relatively hidden animals as the heart-urchins in their snug burrows possess these organs, so probably they too are not spared the menace of settling larvae. Sometimes, the test of a dead urchin will lie on the sediment at the bottom of a calm sea, an oasis for the settler in an otherwise desert expanse of silt, and the test, when it turns up later in a dredge, will be dotted with sessile animals such as ectoprocts, serpulid worms, etc., and these will certainly have settled after the animal's death, because they often spread across the tubercles and pores and even across mouth, anus and apical region. Such specimens often turn up in fossils too.

It is interesting that some members of another invertebrate phylum, the Ectoprocta, possess modified polypides, the avicularia, with a structure remarkably convergent on that of the pedicellariae; they are also said to have a similar function.

The general anatomy of an echinoid pedicellaria is as follows: a number of articulating blades, usually three, hinged proximally, are borne on a movable stem which is supported along at least part of its length by a calcite rod and along the rest by a jelly-like core of mucus (Fig. 22a). The whole organ is covered by an epithelium which has the usual basi-epithelial nerve plexus in association with it, and this plexus is expanded at the base of the stem into a nerve ring. The blades are operated by two series of antagonistic muscles: the large *adductors*, which always have a smooth component and often a striated one too; and the smaller *abductors*, which always contain only the smooth type.

When the jaws of the pedicellaria are touched on the outside, they open; when they are touched on the inside, in most types they snap shut[93]. From this, one infers that a sensory system on the outside of the jaws stimulates the abductors, while one on the inside stimulates the adductors. Fine vibratile processes have been seen projecting from the surface of the inside of the jaws, and these may be associated with the sensory endings of nerves stimulating the adductors, but no such processes have been detected on the outside of the jaws, even by electron microscopy. One concludes that it may be the general basi-epithelial plexus that is sensitive on the outside; certainly, the opening reflex is a much slower reaction than the closing, and the impression is given that the adductor reflex overrides the abductor in some way.

In echinoids, four main types of pedicellaria have been described, though they may vary enormously within the main

divisions. The largest are the *tridactyles* ('three-fingered') (Fig. 22*f*).
A calcite rod supports about the first two-thirds of the stem, the
head being capable of quite a bit of flexure about its neck. The
blades touch at their tips, where there are two or three tooth-like
projections, and the blades are moved by adductors which stretch
between adjacent blades just distal to the hinge, making a triang-
ular pattern when viewed in transverse section. Modifications of
this type, with anything from two to five blades, are found in
various groups. The second type is the *ophiocephalous* ('snake-
headed') (Fig. 22*d*), which are usually the commonest. The chief
features of this type are, first, that serrations are present down the
whole length of the distal part of each jaw, and, secondly, that
each jaw has an inwardly directed process proximally, possibly
to give better holding power. The third type is the *trifoliate*
('three-leaved') (Fig. 22*e*) with broad, leaf-like blades which meet
only along their lateral edges, and a highly flexible, sinuous neck.
They are very much smaller than the other types. The last type,
gemmiform or *glandular* (Fig. 22*b*, *c*) has a calcite rod supporting
the whole stem, has blades which are provided with a tooth distally
to pierce the prey and has poison glands to paralyse it. Round
each gland there is a muscle sheath to bring about discharge of the
poison. On the inside of each blade is a tiny sensory hillock, and
vibratile, sensory cilia cover not only this hillock but the rest of
the inside of the jaws too.

The behaviour of the glandular type is interesting. Like the
other types, they will open if stroked on the outside of the jaws

Fig. 22 Pedicellariae

a Section of a typical pedicellaria, based on an echinoid tridactyle.

b to *f*, echinoid pedicellariae.
b Glandular type, open, and *c* closed, in section, showing poison
glands.
d Ophiocephalous.
e Trifoliate, open.
f Tridactyle, closed.

g to *l*, asteroid pedicellariae.
g Straight type, showing muscles.
h Crossed type.
i Sessile type.
j Pectinate type.
k Alveolar type.
l Valvate type.

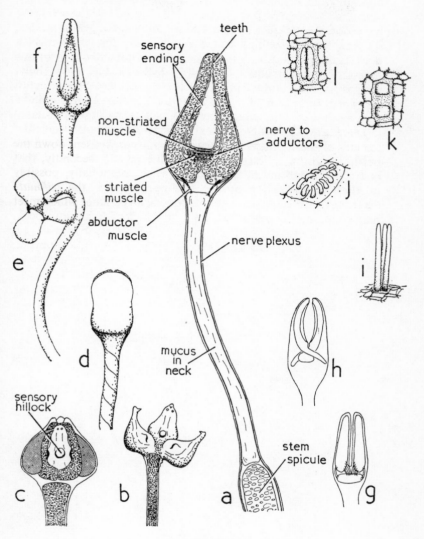

Fig. 22

but will open further if touched with a probe anywhere on the inside. If, however, a piece of a foreign animal, such as a starfish tube-foot, is touched on to the sensory hillock, the jaws will snap shut and the teeth will pierce the foreign tissue; up to one minute later, the poison glands will discharge their contents, though not necessarily simultaneously. Curiously, a chemical stimulus alone, such as flooding with the juice of a starfish, will not evoke closure and poison emission: it is as though the action is brought about by a mechanical stimulation 'primed' by a chemical one.

There are apparently no experiments yet reported on the toxicity of these organs to animals they are likely to encounter, but it is said that if forty of the glandular pedicellariae are boiled in 1 cc. of water and injected into a rabbit, death follows in two or three minutes; the author of this piece of information[94] describes how he narrowly escaped death after touching his native urchin, *Toxopneustes*.

THE TUBE-FEET AND THEIR EVOLUTION

Of all the appendages borne by the echinoderms the tube-feet are the only ones which are found in every living class, and probably were present on many of the extinct forms as well. In their delicacy and in their multiplicity of function they are fascinating, and in their combined strengths in some situations they are astonishing. Though strikingly parallel organs are found in many other groups of the invertebrates, for instance in the food-collecting tentacles of some sipunculoids, with their muscular 'compensation sacs', yet as hydraulic organs the tube-feet must surely take pride of place in the animal kingdom.

These organs exhibit an amazing variety of structure and function; they may be locomotory, tactile or chemosensory, or feeding, burrow-building or respiratory, the latter probably being the original function. Yet the basic histological plan of all of them is remarkably similar. Externally each has a covering epithelium which is continuous with that covering the rest of the test (Fig. 23a). Normally, an integral part of the epithelium is the nerve plexus which ramifies among the bases of the covering cells; where provision is made in a tube-foot for crinkling the surface during contraction, the epithelium and the underlying nerve plexus fold as one layer. The plexus is thickened on one side of the stem to form the longitudinal tube-foot nerve, and also usually at the distal and proximal ends of the stem to form nerve rings. Lining the lumen of the tube-foot is the coelomic epithelium, continuous with that lining the rest of the water vascular system. It is normally ciliated, and in the stems of most tube-feet the cilia are arranged in two longitudinal bands, beating in opposite directions, so that a circulation of coelomic fluid is maintained in the lumen. Between these two epithelia are two important layers: first, next to the external epithelium, is a layer of connective tissue which forms the main structural framework of the tube-foot and may for this purpose contain embedded spicules; and, secondly, the retractor muscles internal to the connective tissue sheath. In addition to withdrawing the tube-foot, these apparently act also in postural bending: it seems that any sector of the muscle cylinder can be contracted independently of the rest (though so far separate innervation to the various parts of the cylinder has not been shown anatomically), and can pull in opposition to the pressure in the opposite direction of the coelomic fluid in the tube-foot lumen to produce flexure.

Distally the tube-feet may be expanded to form a disk. The most familiar form of this is as an expanded plate forming a sucker. This has evolved, probably independently, in some members of at least three living groups, the asteroids, echinoids and holothuroids. In the other two living groups, ophiuroids and crinoids, the tube-feet are normally papillate with somewhat pointed ends.

Fig. 23   Tube-feet

a Diagrammatic longitudinal section of a suckered tube-foot, based on that of *Echinus*. Skeletal parts and mucous glands densely stippled; epithelial and connective tissues sparsely stippled.

b Food-catching tube-foot of a crinoid, based on that of *Antedon*, showing one of the mucus-producing papillae enlarged.

c and d, asteroid tube-feet.
c Digging tube-foot of a phanerozone.
d Suckered tube-foot of a spinulosan.

e Tube-foot of an ophiuroid, showing the ampulla-like expanded part of the canal.

f to l, echinoid tube-feet. See also a.

f to i, diagrams of the disks in plan view, showing the evolution of the skeletal structures in penicillate tube-feet of spatangoids.
f *Echinus*, g *Brissopsis*, h *Schizaster*, i *Echinocardium*.
j Respiratory tube-foot of a spatangoid.
k Funnel-building tube-foot of a spatangoid, showing the excavating scraper and the mucus-producing papillae.
l Feeding tube-foot of a spatangoid.

m Holothuroid suckered tube-foot.

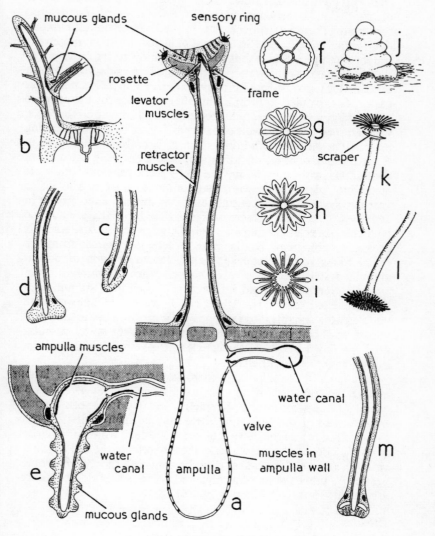

Fig. 23

If a tube-foot is to be extended, some method of forcing coelomic fluid into the lumen is required. In the crinoids this is brought about by isolating parts of the water vascular canal and contracting the compartments thus formed so that fluid is pumped out to the tube-feet[110]. In some ophiuroids, too, a somewhat similar method is found, but here, in addition to the isolation of successive parts of the water vascular canal, the side-branch from the canal to a tube-foot, though embedded in the ambulacral ossicle, is expanded to form a muscular chamber for the purpose of protraction[118]. In asteroids, echinoids and some holothuroids, where rather greater powers of extensibility are seen, blind-ending muscular *ampullae* protrude into the perivisceral coelom (Fig. 23a). Each ampulla consists of little more than a covering epithelium, continuous with that lining the perivisceral coelom, an inner epithelium, continuous with that lining the rest of the water vascular system, and a central sheath of muscle, antagonistic in activity to the stem retractors of the tube-foot; there may also be a thin layer of connective tissue between the muscle and the coelomic epithelium. At the point where the branch from the radial water vascular canal joins the lumen of each tube-foot/ampulla system there is a valve, so that when the muscles of an ampulla contract the fluid in it can pass only into the lumen of the associated tube-foot. This valve, however, is endowed with muscles which enable it to open against a pressure in the tube-foot lumen, so that on retraction of the tube-foot the water vascular fluid can flow back down the canals of the water vascular system as well as into the ampulla.

In almost all echinoderms there is some division of labour among the tube-feet of any one animal: in almost all forms a group of tube-feet round the mouth differ from those arising from the rest of the body, and there may be further divisions in those of the rest of the body. The oral tube-feet, which often arise from the circumoral water ring rather than from the radial water vessels, are nearly always entirely sensory, though in some groups, notably the holothuroids and some echinoids, they may assume an important function in feeding.

Crinoids

Of all the tube-feet of any echinoderm that I have ever seen in activity, those of the comatulid *Antedon* are by far the most vigorous. They are mainly used as feeding organs, with possibly a subsidiary sensory function. The animal sits attached to the substratum by its cirri with its arms held out to form a collecting bowl, its mouth being at the centre. The food-collecting grooves, extending along each arm and pinnule, except the oral pinnules, are lined by tube-feet in groups of three. Along the arms the tube-feet of each group are of equal length, and when extended lie at an angle of about forty-five degrees to the vertical plane of the arm, but along the pinnules the three tube-feet of each group are of different lengths and each lies at a different angle: thus, the largest of each group projects nearly horizontally in the valley between two adjacent lappets (the cover plates which can fold over and protect the groove), the medium-sized ones project at about forty-five degrees and the smallest stick almost vertically upright.

The external epithelium of each tube-foot is drawn out at intervals into finger-like papillae (Fig. 23b). Each papilla is solid, with a single muscle fibre at the centre running its entire length; no other muscles are present in the epithelium. At the base of each papilla are five or more single-celled mucous glands with long ducts passing up the length of the papilla to open at its distal end, and a group of nerve cells which send sensory processes to protrude beyond the end of the papilla. Watching *Antedon* feed, one gains the impression that the tube-feet are stimulated into their violent, flailing activity by the touch of food on the sensory ends of their papillae. These then possibly cause the central muscle to contract, which squeezes mucus from the glands, and at the same time the tube-foot bends violently, the effect of this being to throw out a series of mucous strings in which the food is trapped. Now the task is to get these strings safely into the food grooves, where the cilia will transport them to the mouth. This is where the three sizes of pinnular tube-feet come in: the largest are clearly the main food-catching ones, and these get rid of their food-laden strings by throwing them to the medium-sized ones, which bend both outwards to pick up the strings and inwards to pass them to the smallest ones, and these bend in all directions to push the strings into the food groove. This differential activity is reflected by the retractor musculature of the three sizes of tube-foot: in the largest the muscles are on the oral (uppermost) side of the stem only, in the medium they are on the oral and aboral sides only, while on the smallest they are all round[110].

Asteroids

Two main types of tube-foot are found in this class: pointed in those which burrow (mainly phanerozones) and suckered in those which move over the surface of the substratum. In the pointed ones (Fig. 23c) the connective tissue sheath is expanded at the end into a strong arrow-head for thrusting into the substratum during burrowing, and the epithelium at the tip is heavily charged with mucous glands. These tube-feet probably have a dual function: they assist in feeding by helping to move particles along the ambulacral grooves towards the mouth and they probably also extend up the lateral sides of the arms during burrowing, even possibly across the oral surface, to plaster mucus on to the burrow walls and prevent them falling in.

The connective tissue at the distal end of the suckered tube-feet (Fig. 23d) is also expanded to form the main framework. Smith[116, 117] has shown how the forces exerted by the retractor muscles during adhesion tend to maintain the full diameter of the sucking disk by pulling on the most peripheral of the connective tissue strands. Here again the disk is well supplied with mucous glands, to help in adhesion.

There is an interesting anatomical difference between the burrowers and the non-burrowers: in some of the burrowing asteroids each ampulla has two separate lobes to it, whereas the ampulla of the suckered tube-feet is a simple sac. There are two possible explanations of the bilobed arrangement: first, since each lobe is separately innervated, that only during burrowing activity, when more extension is required, is the second lobe activated; or, secondly and more probably, that a single ampulla has been secondarily subdivided to provide greater surface-to-volume ratio, and hence more muscles in the ampulla wall, so that during burrowing the tube-feet can make powerful thrusts into the substratum.

Echinoids

The recent regular echinoids have the most highly developed suckers on their tube-feet of any class. The sucker disks are supported and assisted in their activity by a fairly complex series of calcareous ossicles, which functionally replace the rather dense connective tissue plate of the asteroid suckered foot[111]. At the distal end of the stem there is a complete ring of ossicles, the *frame*, which lies on the proximal side of a much larger skeletal structure, the *rosette* (Fig. 23*a*, *f*). The function of the frame is to act as an anchor for a special set of muscles. The rosette has two functions: first, it helps to keep the shape and width of the sucker during adhesion, when all the forces acting to raise the diaphragm are tending to pull in the disk edges; and, secondly, it operates during detachment by transmitting the pull of the stem retractor muscles to the edge of the disk, which is the most effective place to apply a pull for releasing the sucker. The stem retractors are inserted at the connective tissue sheath adjacent to the inner border of the rosette to be effective in this way. Within the lumen is another set of muscles, the *levators*, which originate at the level of the skeletal frame and run between the retractor muscle fibres before passing distally to attach at the centre of the disk. Between the rosette and the disk surface are large multicellular mucous glands to help in adhesion, and round the periphery of the disk, at least in the more advanced regular echinoids, is a sensory ring.

So far as one can tell from the fossil record, the dramatic change from regular to irregular, and the consequent potential for exploiting new habitats, occurred more than once: the holectypoids and clypeasteroids arose independently of the spatangoids. The first two orders are included in Zittel's super order Gnathostomata (with a jaw apparatus), which suggests rather less divergence from regular echinoid structure. This is true of tube-foot structure too; though there appears to be no detailed histological study of the tube-feet of any of the three holectypoid species still extant (two species of *Echinoneus* and one of *Micropetalon*), Lovén[56] gives a picture of the skeletal structures in the disk of *E. semilunaris* showing these to be similar to those of regular urchins. In clypeasteroids, too, the tube-foot structure of at least one, *Echinocyamus*, appears to deviate very little from the regular echinoid plan[109], with the exception that the skeletal elements of the disk are missing and are functionally replaced by epithelial muscles which apparently serve to raise the outer parts of the disk relative to the centre during detachment.

So far only the suckered tube-feet occurring over the lower parts of the test have been described. In addition to the difference between these and the wholly sensory group round the mouth (p. 138), there is further division of labour in each ambulacrum in that aborally (i.e. nearer the apical disk) the tube-feet tend to lose their suckers and become, apparently, mainly respiratory and sensory in function.

In the spatangoids, where the tube-foot pattern is drastically changed because of the highly developed burrowing habit of most of them, the demarcation of the solely respiratory tube-feet is much more abrupt. In addition, the suckered tube-feet are no longer required, and they are functionally replaced, in appropriate positions on the animal, by burrow-building, sensory or feeding tube-feet. Thus, taking the common British heart-urchin *Echinocardium* as an example[108], the tube-feet in the dorsal parts of four of the five ambulacra are respiratory, those in the dorsal part of the fifth have the function of building a special respiratory and feeding funnel from the burrow to the surface of the substratum, the feet round the mouth are for feeding only, some on the posterior side build a special sanitary drain, and those round the ambitus (the widest part of the test) and in parts of the oral surface are for sensation only.

The surface area of the disk of mucus-producing tube-feet is normally vastly increased by the presence of numerous papillae (Fig. 23*k*, *l*). The disk in these tube-feet is not supported by a rosette, but each papilla has a single spicule supporting it, at least along part of its length. One can trace a morphological series in the papillae and their skeletal elements from the typical regular echinoid condition (Fig. 23*f*); through the funnel-building tube-feet of such brissid spatangoids as *Brissopsis* (Fig. 23*g*), in which the disk edge is merely scalloped and the fifteen or so spicules form a structure very like a rosette, touching centripetally; through *Schizaster* (Fig. 23*h*), where the papillae are more marked and the spicules reduced; to *Echinocardium* (Fig. 23*i*), where there are two rows of papillae and the spicules are reduced to mere rods and do not touch at their inner ends. In *Brissopsis* and *Echinocardium*, and probably in the others as well, the stem retractors are inserted at the inner border of the skeletal elements, which suggests that they are homologous with the elements of the regular echinoid rosette. As there is no levator muscle system in the spatangoid tube-foot (there being no sucker), there is no need for a skeletal frame as such. But in the funnel-building tube-feet of *Echinocardium*, for instance, and in some others as well, one (or occasionally two) very large spicules may be present on the side of the disk (Fig. 23*k*) in a position roughly corresponding to that of the frame of regulars. It seems most likely that

the purpose of this is to act as a scraper for the walls of the funnel, the sand removed by them during excavation being taken down to the cavity occupied by the animal and passed to the rear of the burrow, where it fills in the space formerly occupied by the animal as it moves forward. A great deal of mucus must be required to plaster the walls of the funnel excavated in this way, so it is not surprising to find that the epithelium covering the disk and papillae of the tube-feet contains numerous mucous glands. Further, the epithelium has muscle fibres running between the glands, most probably for more efficient discharge, and to ensure that the mucus is cast on to the funnel walls and does not remain on the tube-feet. A comparable arrangement is found in those tube-feet of spatangoids which build and maintain the sub-anal sanitary drain, but it is interesting that in the oral feeding tube-feet, superficially similar to the burrow-building ones, the glands are *not* provided with muscles, because the mucus is required to remain on the tube-foot.

Ophiuroids

The tube-feet of the brittle-stars have received far less attention than those of other classes. In general, they show a structure rather similar to that of the crinoids in that most of them are slender and pointed, and the epithelium down the entire length of the stem is raised at intervals into papillae (Fig. 23e), containing mucous glands and sense organs. Since the ophiuroids use their arms for locomotion (p. 53), the tube-feet have an important function in helping the animal to obtain purchase on the substratum; they also function extensively in feeding by helping to pass food along the arms[36].

Holothuroids

These echinoderms, lying on their sides, show typically a division of labour chiefly because of the adoption of a locomotory function by the tube-feet of the three ventral ambulacra and the reduction of those of the other two to small sensory papillae. This, of course, does not hold good for the Apodida (p. 77) and the other burrowing forms. In addition, the oral tube-feet, even in the Apodida, are usually very large relative to the body size, and may even have an extended length equal to half the length of the body. The wide variety of structure in these organs is used extensively by taxonomists, and indicates that a correspondingly wide variety of

feeding methods is employed by the class. Typically, the oral
feeding tube-feet are dendritic and the stems highly muscular. They
apparently obtain mucus 'second hand' from glands inside the
mouth. When the animal is feeding, these organs are extended in a
sweeping motion, apparently scavenging particles of detritus from
its neighbourhood, then they are bent round into the mouth while
retracting somewhat, and finally they wipe their mucus with its
collected food off against the oral ring of ossicles and extend again[66].

The suckered tube-feet (Fig. 23m) show a pattern which is
different again from that of the asteroid and echinoid suckers.
Here there is a single, slightly dome-shaped spicule as a support,
with a ring of very minute arcuate spicules on its proximal side,
peripherally. Groups of unicellular mucous glands lie distal to the
skeletal plate, and connective tissue fibres pass through its
fenestrae to be inserted at the disk cuticle. How the sucker
functions is not known precisely, but it is probable that the
stickiness of the mucus and the deformation of the peripheral
epithelium, relative to the domed spicule, are responsible for
adhesion. The retractor muscles of the tube-foot stem will tend
to raise the domed spicule from the surface to which the tube-foot
is adhering, and the peripheral epithelium, supported by its tiny
arcuate spicules, will form the sides of the sucker. Detachment can
occur only when the retractor muscles pull the sucker at an acute
angle to the surface of the substratum.

The evolution of the tube-foot/ampulla system

So rarely are the remains of tube-feet and associated structures
left behind in the fossil record that we know very little of the
origin and early history of this system. The situation is not helped
by the fact that today's crinozoans, the Crinoidea, have the water
vascular system wholly embedded in the soft tissue wall away from
the skeleton, so that no imprint of the system is made on those
parts likely to fossilise, and one is forced to conclude that the
extinct crinozoans may have had the same arrangement. There is,
however, one piece of indirect evidence that the cystoids, eocrinoids
and paracrinoids may have had a tube-foot system: they have
two accessory pores in the theca, one of which was probably a
hydropore, and if this had the same function as the madreporite
in today's echinoderms, then it is hardly likely to have been
present unless there was an interconnected water vascular system
operating protrusible tube-feet.

The first actual traces of a tube-foot system are found, surprisingly, among the earliest confirmed occurrence of fossil echinoderms, in the Lower Cambrian. In one of the curious spiral helicoplacoids, *Waucobella*, the single ambulacrum shows a column of pores between adjacent ambulacral plates, pores which probably carried canals either from an external water vascular canal to internal ampullae[137], or from an internal canal system to the tube-feet[76]. Which of these two possibilities is correct is not possible to decide on present evidence; but at least we can say that these very early echinoderms almost certainly had tube-feet.

Among the edrioasteroidea, which also occur in the Lower Cambrian rocks, the picture is by no means clear. The earliest forms, agelacrinids such as *Stromatocystites* (p. 101), show no evidence of ambulacral pores, and though some writers[75] suggest that the shape of the ambulacral cover plates is consistent with their having had soft structures protruding between them, this does not appear likely. Yet later edrioasteroids, such as *Edrioaster* (p. 102), undoubtedly had ambulacral pores which almost certainly carried canals to internal ampullae[73].

The Ophiocistioidea (p. 106) had tube-feet without a shadow of doubt, because the organs had large embedded skeletal plates which are preserved in the fossils. The Cyclocystoidea (p. 109), too, had large peripheral cups, supplied by a canal from the interior of the theca, which most likely bore tube-feet. Lastly, the structure of the homalozoan feeding arm strongly suggests that it bore a typical echinoderm ambulacrum, with tube-feet having the power to extend and contract in an ophiuroid-like manner (p. 115).

So the evidence suggests that tube-feet may well have been present in all the echinoderm classes in which the structure is reasonably well known.

A helpful evolutionary sequence can be established from the comparative morphology of the water vascular system in modern classes. We have seen (p. 138) that in the crinoids it is the radial water vascular canal that creates the necessary hydrostatic pressure for protraction of the tube-feet. In ophiuroids, the situation is slightly more advanced, for here, in addition to the radial canals playing their part in the process by elastic contraction, the lateral canals leading to the tube-feet also contract. In the next stage, shown by the asteroids, not only do the radial and lateral water vascular canals contribute to tube-foot protraction, but accessory ampullae are present as 'offshoots' of the lateral canal. In this case, contraction of radial and lateral canals produces about half the extension of each tube-foot: the final half is brought about by the activity of the ampulla. It follows from this that the expanded volume of each ampulla is about half the expanded volume of the tube-foot lumen; it follows also that the valve isolating each tube-foot/ampulla system from the rest of the water vascular system must be able to open or shut against a pressure on either side of it, and this does appear to be so (p. 138).

The situation in echinoids is somewhat similar. Here, too, accessory ampullae are present in the system, though the actual mechanics of the operation have not been worked out. In some holothuroids, such as *Holothuria* itself, there are no accessory ampullae, and therefore the process again depends on the generation of pressure within the canals themselves. But in others, such as *Cucumaria*, ampullae do occur. Here again, the system has not been worked out.

So far as the evolution of the tube-feet themselves is concerned, the more primitive groups had finger-like structures, well endowed with mucous glands over their entire surface, and these were well suited to a feeding function (as in crinoids) or in burrowing activity (as in primitive asteroids and ophiuroids). For surface dwelling, however, the tube-feet required better adhesive powers than mucus alone could bestow, and so the sucker-disk arose (as in later asteroids, echinoids and holothuroids). Subsequent return to burrowing (as in some irregular echinoids) meant a return to the mucus-producing organs, and the secondary loss of the sucker.

LARVAL FORMS AND THEIR METAMORPHOSIS

Whilst radial symmetry is one of the characteristic features in the adult echinoderm and an attached condition is very widespread, in the larval state the various classes share a bilateral plan, and are invariably free for most of their larval life. It may be that the extinct groups also developed indirectly, through larval forms which were probably very similar to those of living echinoderms. A larva has a double task to perform: to distribute the species and to grow up into the adult. In the echinoderms, as much as in any other animal group, we can see that larval evolution has taken place independently of that of the adult, and is subject to conspicuous specialisation, even among closely related types; for example, among the echinoids *Eucidaris* has a planktotrophic larva (feeding on plankton), a free-floating form with long arms, whereas *Heliocidaris* has a lecithotrophic larva (feeding on stored yolk), which has the shape of a barrel[126]. Some recent groups even dispense with a larval stage altogether: that is, they show direct development, usually but not always coupled with the possession of rather more yolky eggs and with some form of marsupial or incubatory care of young by the parent (p. 156); but this is the exception, at least in temperate echinoderms; in most cases the eggs and sperm are shed into the sea, and fertilisation and development take place there.

Early development
The egg of most echinoderms is a sphere of about 75 μ in diameter surrounded by a jelly coat; the sperm is about 50 μ in length, with

a pointed conical head, a middle-piece the shape of a flattened
cylinder containing two centrioles, and a long thin tail. There is
an acrosome, like an arrowhead, at the tip, and it has been said
that there is also a pore through which a sticky substance is
secreted as the sperm meets the egg.

Cleavage of the echinoderm egg is total, indeterminate and
radial. The resulting blastula is oval, hollow and ciliated all over.
An impushing at one end forms the beginnings of the larval gut,
or *archenteron*. The hole through which the archenteron opens to
the exterior is the *blastopore*. The blastopore marks the position of
the future anus: that is, the echinoderms are *deuterostomes*. There
now follows a stage of development which is considered highly
important from the point of view of phyletic relationships:
coelom formation. It is also a stage in which considerable variation
occurs, which makes it all the more difficult to interpret. In the
vast majority of the echinoderms it happens by *enterocoely*; that
is, the coelomic sacs begin as outpushings from the wall of the

Fig. 24   Larval types in asterozoans and echinozoans
    The asteroid, holothuroid, ophiuroid and echinoid larval types can
all be traced from a basic dipleurula. Thick lines are ciliated bands.
a Dipleurula, side view.

b to *f*, asteroid larvae.

b Early *bipinnaria*, ciliated band divided into two.
c Late *bipinnaria*.
d *Brachiolaria*, with adhesive disks at anterior end.
e Attached *brachiolaria*, star rudiment breaking off.
f Young star.

g to *j*, holothuroid larvae.
g *Auricularia*.
h *Doliolaria*.
i *Pentactula*.
j Young cucumber.

k to *n*, ophiuroid larvae.
k Early *pluteus* with four arms.
l *Ophiopluteus* with eight arms, one pair of epaulettes. Spicules
    omitted.
m Late *ophiopluteus* from front.
n Young brittle-star.

k to *q*, echinoid larvae.
o *Echinopluteus* with twelve arms, two pairs of epaulettes.
p Late *echinopluteus*, showing similarity to late ophiopluteus (*m*).
q Young urchin.

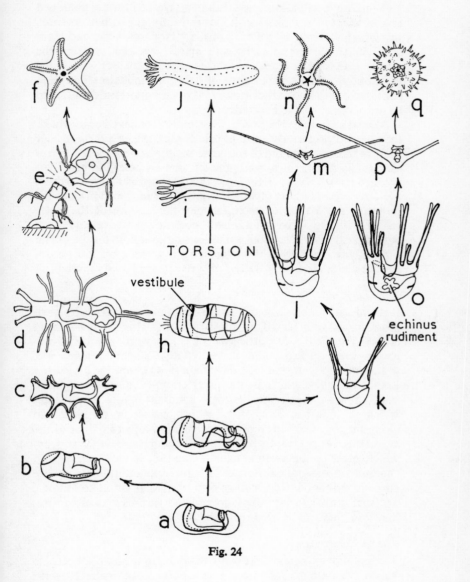

TORSION

vestibule

echinus
rudiment

Fig. 24

archenteron, one on each side. But in a very few forms, notably a few species of ophiuroids, it develops by the other method, *schizocoely*; that is, by splitting of the mesoderm, a method more usual in those invertebrates with spiral cleavage, such as the annelids. This latter type of cleavage is also the method seen in the vertebrates, and its occurrence in a few echinoderms has been considered by some authorities[143, 144] to be important with regard to the ancestry of the chordates.

Almost as soon as the primary coelomic sacs have been formed, they bud off posteriorly another pair of sacs, the *somatocoels*, later to form the main coelom of the adult body; then the right anterior sac usually degenerates, while the left one becomes bilobed. This is called the left *axohydrocoel*, the anterior part being an unpaired, pre-oral *axocoel* (most of which degenerates in the adult) and the posterior the *hydrocoel* (later forming the cavities of the water vascular system). The left axohydrocoel now forms a connexion with the exterior by a small canal leading to a hole on the dorsal surface, the *hydropore*. Meanwhile, an inpushing on the ventral surface pierces the body wall to give rise to the larval mouth.

The dipleurula

It is understandable that early in life a common plan exists between those forms with larvae, or most of them. This stage has been termed the *dipleurula* ('little two sides'), a word which has unhappily been used to denote also a hypothetical common ancestor to the entire phylum (p. 167). The main thing about the dipleurula is that the cilia, which in the early embryo covered the entire body, now aggregate into a single band forming a closed loop which starts just in front of the mouth, passes up the sides and along the top, then down and under the larva just in front of the anus (Fig. 24a). This, then, represents the basic larval type from which most of the various later larvae are derived. No known echinoderm has only a dipleurula before metamorphosis: all are subsequently modified for their own particular larval conditions. We shall now consider the larvae of the various classes in turn, and how they are derived from the basic pattern.

The holothuroids

The simplest situation is probably that seen in the holothuroids, in which the dipleurula is transformed into an *auricularia* by the ciliated band becoming sinuous (Fig. 24g). In this case the complement of coelomic sacs is the same as in the dipleurula, namely, left axohydrocoel, left and right somatocoels. Before metamorphosis into the adult, the sinuous ciliated band reorganises by breakage, etc., into three or four rings round the barrel-shaped body (Fig. 24*h*); the larva is now called a *doliolaria*. In some forms, such as *Cucumaria planci* and *Leptosynapta inhaerens*, the auricularia stage is omitted, while in a few other species of *Cucumaria*, e.g. *saxicola* and *frondosa*, the egg develops into another sort of larva altogether, which has cilia all over it, not restricted to bands. As in other echinoderms, some members of the holothuroids brood their young, but this does not necessarily mean that they dispense with a larval stage: in *Chiridota* a doliolaria stage is passed through by the embryo while incarcerated in the parent's coelom.

In the doliolaria the beginning of the gut, just inside the slit-like mouth, has a thickened wall, and this is called the *vestibule*, a structure which is important in the later development of the oral region. The middle part of the axohydrocoel forms a five-lobed ring round the vestibule, destined to form the five primary oral tube-feet, followed by a sixth lobe, later becoming the polian vesicle; five canals are then budded off, which will become radial water vascular canals. The opening of the hydropore now disappears in those forms with no madreporite in the adult. Five cavities, destined to become the epineural sinuses, originate by inpushing of the vestibule wall, and at the same time a ring-like mass of nervous tissue, the future oral nerve ring, also arises from the vestibule wall. The vestibule and the structures which are forming from it now move towards the anterior end of the larva, and the vestibule wall becomes the buccal membrane. At this stage the blastopore closes off, but the anus opens again a little way from it, a curious but fairly common phenomenon of development in deuterostomes which is not fully understood. The lobes of the hydrocoel now push through the buccal membrane and form the lumina of the five primary buccal tube-feet. Usually at about this time, too, two or three tube-feet, the terminal tentacles of the trivium, are formed near the posterior end of the body. The larva at this stage is called a *pentactula* (Fig. 24*i*) and now it settles on the bottom. The adult form is gradually acquired by the growth of other tube-feet over the body and the development of internal organs, such as the respiratory trees from the cloaca. This is a necessarily brief and simplified account of holothuroid development; in general, however, it can be seen that the main parts of the larva are retained in the adult; this is not true of some of the other groups, such as the next.

The asteroids

Very early in larval life in this group the single ciliated band of the dipleurula divides into two, a small pre-oral band and a larger post-oral one. This larva is called the *bipinnaria* (Fig. 24*b*). The development of the asteroid coelom varies somewhat from the general pattern described above. The original pair of pouches, in addition to budding off a pair of somatocoels posteriorly, grow forward and fuse in front to form a U-shaped cavity enclosing the front part of the gut. Most of the accessory coelomic structures, that is, cavities other than the ordinary body (perivisceral) cavity, derive from the left side of this sac, as is normal, but some structures may form from the right side, as will be mentioned later. Both ciliated bands become sinuous, and in some asteroids, such as some species of *Asterina* and *Astropecten*, metamorphosis occurs without further change; but in others, such as *Asterias*, the sides of the larva become drawn out into a number of processes, the contours of which are followed by the ciliated bands (Fig. 24*c*). Next, in most asteroids, the anterior end of the larva becomes drawn out into three arms, each provided with a terminal adhesive disk and with a fourth between their bases; extensions of the axocoel pass into these arms. This larva is termed a *brachiolaria* (Fig. 24*d*). The subsequent development of coelomic and accessory structures has been well worked out[37]. Suffice it to say here that the central part of the left axohydrocoel buds off a five-rayed ring, later to become the circum-oral water ring and the lumina of the five primary tube-feet, and a stone canal develops from this ring to the hydropore. At this stage the remains of the right axohydrocoel in some forms are said to give rise to the dorsal sac, though in others this structure is thought to arise from mesenchyme at the anterior end of the body, while in still others it is said to arise by invagination of ectoderm; the sac has been seen to undergo pulsations, apparently to maintain circulation in the larva.

After about two months at the brachiolaria stage the larva attaches by its adhesive organs, and after a further short time the young starfish which has been forming on the left side of the larva breaks free by a rupture which cuts off the larval mouth from the rest of the gut (Fig. 24*e*). A new mouth pushes in through the water ring and joins up with the larval stomach. Similarly, a new anus is formed opposite the new mouth, that is, on the right side of the larva, which is now the aboral side of the young starfish. There has been a change of symmetry at metamorphosis from bilateral to radial.

The echinoids

Next in complexity come the *pluteus* larvae; this term, originally coined by Müller in the 1850's, means 'easel', but unhappily this refers to the appearance of the larva when upside down! Both echinoids and ophiuroids have this type of larva, distinguished by the appropriate prefix, i.e. *echino*pluteus and *ophio*pluteus. The main feature of the plutei is the great extension of the larval arms, the legs of the easel; these are supported by calcareous spicules (p. 123), later discarded. In the early echinopluteus two pairs of arms form first, one pair in front of the mouth and the other in front of the anus (Fig. 24*k*), then a further four pairs appear between them (Fig. 24*o*). The main ciliary band (single) follows the contours of the arms, but two ciliary patches, the *epaulettes*, appear at each end of the larval body; these are said to be the main locomotory organs of the larva. Sometimes, particularly in the irregulars, a process, the *spike*, projects downwards from the aboral side, possibly to keep the larva upright.

The process of metamorphosis has been carefully worked out for numerous species. In brief, the common features of it are as follows: the left axohydrocoel forms a five-lobed sac on the left side of the gut, and adjacent to this an inpushing of the left side of the larval body occurs, forming a vestibule, the entrance of which later closes over. Inside the vestibule the primary podia grow out from the lobed hydrocoel, the epineural canals fold in from the wall and the nerve ring also forms from the wall, as in holothuroids. Meanwhile, the left somatocoel has surrounded the hydrocoel and produced buds between the lobes which will eventually give rise to the elements of Aristotle's lantern. Calcareous elements begin to be deposited[89]. The developing material is now concentrated into a spherical disk on the left side of the larval gut, a disk which the original describers called the *echinus rudiment*; this name has been retained, even for other genera. At metamorphosis the vestibule wall breaks open so that the primary podia can emerge, the coverings of the larval arms are resorbed and their spicules discarded, and a new mouth and anus are formed on the left and right side of the larval body respectively. After the break-through the young urchin, 1 mm or so in diameter, sinks to the bottom.

The ophiuroids

The *ophiopluteus* develops from the dipleurula stage in very much the same way as the echinopluteus, but has only four (sometimes three) pairs of arms finally (Fig. 24*l*), all with skeletal supporting rods. One pair of swimming epaulettes may also appear. Internal development, however, does not follow the echinoid plan. The hydrocoel, instead of developing on the left side of the gut, forms a ring round the oesophagus, and the larval mouth forms the adult mouth; the larval anus closes over, there being no anus in the adult. No vestibule is formed, though the primary podia, when they develop from the hydrocoel ring, emerge into the gut near the mouth, which is equivalent in position to a vestibule. The nerve ring develops simply as an ectodermal swelling and becomes covered by an epineural sinus through the overgrowth of folds of epidermal tissue. The larva continues to swim around during metamorphosis, the larval arms are resorbed and the rods discarded, as in echinoids, and calcareous elements laid down in the body. Then, as the larva settles, the future arms start to grow out from its sides.

The crinoids

This group remains somewhat apart from the others because the gastrula develops direct into a *doliolaria* (Fig. 25*a*), not unlike that of the holothuroids. This has four or five bands of cilia and an adhesive disk somewhere on or near the first band; there is an apical sensory tuft of cilia at the front end. An archenteron forms by invagination but here the blastopore closes off completely, so that the archenteron is now a closed sac lying in the blastocoel, that is, the cavity of the blastula. The archenteron eventually divides into two somatocoels, a hydrocoel, and axocoel and an enteron. There now appears a depression somewhere between the first and third ciliated bands, which will become the vestibule.

All the above structures are formed while the embryo is still in the egg membrane; it now escapes, and remains in the plankton for a matter of hours, or at most days, before selecting a suitable spot with its apical tuft and settling on its adhesive disk (Fig. 25*b*). The vestibule, by now quite a large sac, becomes cut off and the

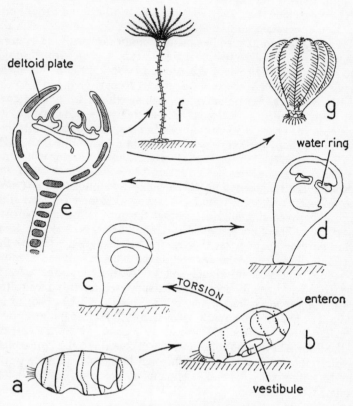

Fig. 25 Larval development in crinozoans

Crinoids do not pass through a dipleurula stage.
a *Doliolaria*, with enteron already cut off inside.
b Settled *doliolaria*, vestible enlarging.
c *Cystidean* after torsion of vestibule.
d Late *cystidean*, water ring surrounding future mouth.
e *Pentacrinule*, with tube-feet forming, and gut growing round towards left side.
f Adult fixed crinoid.
g Adult comatulid, after breaking larval stem.

whole inner mass rotates through ninety degrees, bringing the vestibule from the ventral side of the larval body to the future oral (upper) side (Fig. 25c); between the vestibule and the enteron the hydrocoel forms a five-lobed ring and then connects with the axocoel, which in its turn has made a connexion with the exterior by a hydropore. The vestibule now sends a projection through the hydrocoel ring to join the enteron (Fig. 25d) and this is the future mouth, while the lobes of the hydrocoel grow up into the vestibule as the forerunners of the primary tube-feet. At this stage the larva is called a *cystidean*, because of its supposed similarity to a cystoid. There is no anus as yet, but the gut grows round from the right side of the enteron to the left in a loop (Fig. 25e), later to pierce the side of the larva and form the anus, but this will move to the oral surface as development proceeds. The roof of the vestibule is strengthened at this stage by five interradial spicules, the *deltoids*, and soon radial grooves, which later become slits, appear between the deltoids, so that the roof is made up of five short arms. These are not the future adult arms, because the adult arms develop from a second, radial series of plates below the vestibule. The deltoids now bend upwards to permit the primary tube-feet, and later others, to emerge. At this stage the larva is termed a *pentacrinule*, because of its supposed resemblance to the fossil *Pentacrinites*. The plates of the theca are laid down in its walls. The hydropore now closes over, but to replace it many pores develop in the hydrocoel (now water vascular) ring, and others in the floor of the vestibule (now the tegmen). In the comatulids several months may be passed through in the pentacrinule stage, but then the animal breaks its stem at the top and thereafter leads a free existence.

Some general points on development

As has been emphasised, most echinoderms take no care of their young; but a few members of all classes brood them, either in pouches within the body, e.g. the genital bursae of ophiuroids or the stomach of some starfish, such as *Leptasterias*[121] (feeding ceases while this is going on), or in accessory structures outside, e.g. the posterior respiratory ambulacra of the irregular echinoid *Abatus*, which are particularly deep for the purpose. Many other brooding methods, too numerous to mention here, are shown by members of the phylum.

Comparison of larval development has been used quite extensively to suggest or support phyletic connexions[7, 128]. It is unfortunate that the crinoids, on other grounds considered primitive among the living groups, have omitted the otherwise common early stage of larval life, the dipleurula. The holothuroids share a doliolaria larva with the crinoids, and since, first, some holothuroids also reach the doliolaria stage directly, and, secondly, both groups develop a vestibule which migrates by torsion to the future oral side (they are the only groups to do this), the similarities may appear to indicate relationship. But, on the other hand, in the holothuroids the future water vascular ring grows round the vestibule, while in the crinoids it forms to one side of it, the future mouth pushing through to form the ring. And on comparative anatomical grounds we would not say that the crinoids and holothuroids are close; unfortunately, the fossil record cannot help, but in general it seems to be agreed that their similarities of larval type are convergent.

The rather primitive position of the asteroids, suggested by some features of their adult anatomy (p. 162), is thought to be confirmed by the fact that the brachiolaria larva attaches by its anterior end for the latter part of its life in very much the same way as the cystidean of crinoids; one would expect a difference in subsequent development, however close they are phyletically, because of the different positions of the mouth, and this is shown by the fact that in crinoids there is torsion to bring the mouth uppermost, and in the asteroids there is breakage so that the starfish can settle mouth downwards.

Of more general phyletic interest is the nature of the coelom. The presence of a basic complement of five sacs, that is, an unpaired, pre-oral axocoel and paired hydrocoels and somatocoels, suggests a possible connexion between echinoderms and other coelomate phyla with such a provision of coelomic compartments, such as the chaetognaths, pogonophorans and hemichordates. Of these, the latter group seems to be the closest relative[133], principally because some have tentacles with coelomic cavities from the central sacs (hydrocoel of the echinoderms, mesocoel of the hemichordates).

One very interesting common feature is possessed by all early echinoderm larvae: the hydropore. Almost nothing is known about its function in the larva; it has been thought of as an exit for accumulated waste from the coelomic pouches—a sort of metanephridium, in fact—or as a pressure equaliser, but it

connects only the left axohydrocoel with the exterior, the somatocoels remaining closed, except for the occasional temporary
canal to one of the hydrocoels. Possibly the hydropore is so
important for the efficient operation of even the primary tube-
feet, which become operational very soon after metamorphosis,
that it needs to precede their appearance. When the body of the
post-larval echinoderm is rigidly supported by spicules the
operation of the tube-feet would be seriously hampered by
changes in the external water-pressure, e.g., by tidal rise and fall,
as has been suggested previously (p. 40). More will need to be
known about the organisation of the larva and the mechanical
problems of its development before we can hope to be nearer a
solution to this problem.

Another common developmental feature of the phylum is the
arrangement of those body plates which form first. This follows
a fairly constant pattern in all those with an extensive skeleton,
that is, all living classes except holothuroids. Usually the first
plates to form are a ring of five interradial plates (*genitals* of
echinoids, *interradials* of asteroids, *basals* of crinoids), a ring of
five radial plates outside and between them, and a central plate
(*centro-dorsal* of crinoids, *central* of asteroids, *suranal* of echinoids). Then, in the stellate forms a further radial ring, the
terminals, is laid down. There has been discussion[32, 40] on whether
the common arrangement of these plates is the result of homology
or merely convergence. The arguments are beyond the scope
of this book, and in any case they are somewhat harassed by
the total lack of very young specimens from the fossil record,
specimens which might well show how and in what relationship
these plates are formed, and confirm or deny their homology.
Suffice it to say here that the similarity in the arrangement of the
aboral (apical) plates is striking, and it may be that the constant
arrangement has something to do with the determination of
pentamery in the group, as described in Chapter 10.

14

THE PHYLOGENY OF THE ECHINODERMATA

The unfortunate thing about building a phyletic tree of any animal phylum is that one can never trust the early fossil record: though it is perfectly true that only a sequence of fossils can indicate the actual course of evolution, yet the vagaries of fossilisation may have caused primitive groups to appear in the record later than less primitive ones. We have only to recall, for instance, that the monoplacophoran molluscs are unknown from rocks between the Cambrian period and the present day to realise what a vast span of time can elapse in which no record of a phyletic sequence is preserved. So the time of appearance of a particular class of the echinoderms is not necessarily a reliable guide to its phyletic relationship to other groups and does not indicate whether it is 'primitive' or 'advanced'.

If what we surmise about the invertebrate relations of the echinoderms (see Chapters 1 and 15) is anywhere near the truth, then the classes that would appear to be closest to the ancestral echinoderm, on morphological grounds, are the eocrinoids, crinoids and cystoids, since their members have upwardly-directed mouths, recurved gut and a food-collecting lophophore. Which of these three is the more primitive is, at the moment, anybody's guess, unless we invoke the time-of-appearance argument.

In that case, the EOCRINOIDEA have it (Fig. 26), and indeed their possession of brachioles rather than arms, the fact that they have a complete theca rather than a capsule and tegmen, and the presence of thecal pores bearing epispires may all be primitive features, though it must be stressed that the evidence is far from unequivocal. It seems likely that the CYSTOIDEA, PARACRINOIDEA, true CRINOIDEA, the PARABLASTOIDEA and BLASTOIDEA are all somehow related, since they share the same basic body pattern and differ in details of body plating, respiratory structures and arrangement of the ambulacra.

How the acquisition of five-rayed symmetry fits in is not clear. The eocrinoids were pentamerous from the start, though the cystoids were not. Then two other non-pentamerous groups appear early in the record: the HELICOPLACOIDEA, at the very start of the record, and the HOMALOZOA slightly later. These seem to represent unsuccessful excursions into asymmetry, the one making use of an expandable spiral arrangement for emerging from a protective burrow, the other becoming flattened to sit on the sea-bottom. Both share a single ambulacrum; but related or not, their origin is entirely unknown.

There are reasons for thinking that the EDRIOASTEROIDEA may have arisen from a crinoid-like form, by losing the arms and relying for feeding solely on the thecal food grooves. It should be recalled that in the crinoids a spicule-strengthened tegmen covered the oral surface of the theca, with the mouth at or near the centre. If one imagines a primitive crinoid, such as *Stephanocrinus*, stripped of its arms and pinnules, so that only those parts of the ambulacral grooves on the tegmen are left, then one has a structure not far removed from the basic pattern of such edrioasteroids as *Stromatocystites* (Fig. 17c).

So far, all echinoderms we have mentioned have fed entirely on the rain of detrital matter falling to the sea-floor from the waters above; they have been almost entirely rooted to one spot, their success depending on their ability to command a catchment area sufficient for their needs. But now, while the crinoids continue to exploit this method, a dramatic change occurs in at least one of the lines from them, which involves turning the animal completely upside down, so that the prehensile, and probably by now muscular, tube-feet can be used for locomotion. No longer is the animal rooted, plant-like, to one spot, but is at last eleutherozoic, free to wander over the substratum, with the consequent ability to exploit fresh ecological niches. This means, of course, that the problem of feeding must be solved anew, and one can say that each of the four living non-crinoid classes have solved the problem in a different way.

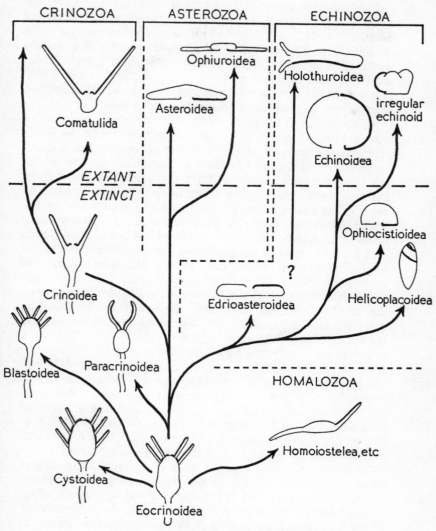

Fig. 26   Radiation of tne echinoderm classes

This diagram shows one possible phylogenetic tree of the echinoderms, and the arrangement of the ambulacra (thick lines) in the
various classes.

First came the ASTEROIDEA, at some time in the Ordovician, and there are good grounds for thinking that they evolved from biserial crinoids, that is, those with a double column of plates in each arm. The crinoid brachial plates became the asteroid ambulacrals and the crinoid pinnulars became the virgalia. Then, we know from the excellent fossil record of the Ordovician that the OPHIUROIDEA arose from the asteroids, though they evolved tantalisingly parallel to the echinoids in some features.

Of the origin of the great group ECHINOIDEA very little can be said. It looks as though the edrioasteroids may have given rise to them, though we have only the evidence of comparative anatomy to go on; one can point to certain edrioasteroids and say that all the transformation necessary is for the animal to break loose, extend its ambulacra and turn upside down to become an echinoid, but one has not explained the migration of the anus and madreporite, and similar problems. Because several of the early Palaeozoic forms had flexible tests and are seldom found preserved in entirety, one wonders whether all the earliest echinoids did not have similar tests, which may explain the total absence of transition forms.

This is undoubtedly the reason why we must admit ignorance on the origin of the last group too, the HOLOTHUROIDEA. Again, very little can be said about this, beyond pointing to the fact that the bodies of early echinoids and holothuroids were both flexible, and thus the two groups may have shared ancestry. Perhaps the similarity of the ambulacral plating of some edrioasteroids with the oesophageal ring of some holothuroids is significant; the possibility of a link on these grounds has been pointed out (p. 79). Their closed ambulacrum tells us very little, because we have seen that such a feature has developed independently in two groups already (ophiuroids and echinoids), in both of which one can trace the actual process through the fossil record.

So despite the excellence of their fossil record the various classes of the echinoderms do not readily reveal their ancestry. Except for the separation of the asteroids and ophiuroids, which in this book are regarded as separate classes, the differentiation of the other classes took place in Precambrian times and so the likelihood of the picture being filled in by future finds remains slight.

15

RELATIONS OF ECHINODERMS
TO OTHER PHYLA

Invertebrate relations and origin of the Echinodermata

To put the echinoderms into perspective in the animal kingdom
we need to trace the history of the invertebrates, as far as can be
done with our limited knowledge, from quite early times, at the
grade when the annelid worms arose from their forbears. The
radiation into the main phyla, however, occurred in the Precambrian, long before we can expect to find much fossil evidence.
Clearly, there are various ways in which data derived from the
study of such evidence as is available to us from present-day
representatives can be arranged so as to give a plausible picture
of evolutionary relationship. What has to be decided is which
arrangement is the most satisfactory in using the greatest number
or relevant facts, at the same time leaving the fewest unaccounted
for. One of the most thought-provoking problems in this kind of
phyletic speculation is the relative weighting one can give to the
various types of evidence. Is one to consider developmental
similarity as necessarily more valuable in tracing descent than
anatomical? Which anatomical features in an animal are least
likely to be influenced by its specific environment and which can
be expected to be reasonably static for our purpose? And so on.
What I have attempted to do in the pages that follow is to present
what I consider to be the most plausible arrangement of the
phyla adjacent to the echinoderms in the phyletic tree, together
with the most important evidence. Of necessity, much is left out,
but anybody wishing to trace this fascinating subject further can
do so through the references quoted.

Now to return to the annelid grade of invertebrate organisation: this was the stage at which metameric segmentation was seized upon for the first time, probably because of the increased locomotor efficiency it bestowed. It was also a stage at which a peculiar form of development appeared, a type of cleavage which though basically similar to the spiral pattern possessed by some of the earlier invertebrates, the platyhelminths and aschelminths, came to show a recognisably unique cross-like pattern of cells on the dorsal surface of the blastula which was different from anything previously seen. This pattern is known as the *annelid cross*. Subsequent development in the annelids leads to a larva called the *trochophore*, which exhibits a fairly constant pattern among all the annelids, that is, with much less variation of form than is found in the larvae of the echinoderms. The larva is roughly spherical, with an equatorial ring of cilia and an apical sense organ, a mouth on one side just below the ciliated ring, and an anus near the base.

Until recently, textbooks placed the Sipunculida as a class of the annelids, principally because they too show spiral cleavage, have the peculiar annelid-type blastula with its cross, and disperse by means of a trochophore larva not unlike that of the annelids. But, on the other hand, there is no trace of segmentation in the adult or in the larva, and they have a recurved gut, a feature possessed by no annelid, even those burrowing forms in which one would expect such an arrangement to be an advantage. Other aspects of their adult anatomy also belie such a close connexion, particularly the possession of a peculiar tentacle arrangement round the mouth. This varies enormously in different sipunculids, but its basic anatomy may well hold a clue to the subsequent evolution of invertebrate phyla and particularly the origin of the echinoderm tubular coelomic system. In general, it consists of a number of tentaculated outgrowths of the body wall, heavily ciliated externally, which provide a catchment funnel for food. The cavities of the tentacles lead into a circum-oesophageal vessel close to the mouth, to which one or more vesicles, the so-called 'compensation sacs', are connected, hanging in the body cavity alongside the gut. The walls of these sacs are muscular, and by their contraction, as one would expect, the tentacles round the mouth are extended. It should be quite evident from this brief account of Sipunculida that in their basic adult anatomy they closely resemble the picture of a primitive adult echinoderm outlined in Chapter 1. But this is by no means the whole story: one expects a good deal of convergence when two groups adopt similar modes of life, and there are features which do not quite fit the picture.

The blastopore of the sipunculid larva is the site of, or close to, the future adult mouth, that is, the phylum is *protostomatous*; but in the echinoderms (p. 148) the blastopore becomes the adult anus (*deuterostomatous*), and these differences in mode of development may represent divergence at an earlier stage in evolutionary history. A protostomatous condition is retained in a line of minor coelomate phyla sometimes misleadingly lumped together as the *Lophophorata*. These phyla are the Phoronida, Ectoprocta and Brachiopoda. All are characterised by the presence in the adult of a hollow-tentacled feeding funnel round the mouth, the *lophophore*, the cavities of whose tentacles are coelomic and connected by a ring or loop round the oesophagus—an arrangement very similar, in fact, to that in sipunculids. Furthermore, the lophophorates have trochophore larvae, metanephridia and, in general, a schizocoelous coelom (body cavity formed by splitting of the mesoderm, though this is secondarily variable), all of which they share with the sipunculids. So one can fairly safely deduce a connexion between the Sipunculida and the Lophophorata[137].

There is some evidence that the lophophorate body is subdivided into three regions: protosome, mesosome and metasome, though the protosome is generally very much reduced. The lophophore itself is borne by the mesosome, and this cavity is quite separate from the metasome. Such division of the body is also shown by phyla on what is almost certainly a second line derived from the sipunculids, including the Chaetognatha, Pogonophora and Hemichordata, and, of course, the Echinodermata (p. 157). Further, in the hemichordates and echinoderms there are tentacled outgrowths from the middle body division, as in the lophophorates; in the echinoderms the ambulacral system of tubes represents this.

Of these four phyla, certain features place the Chaetognatha, the arrow-worms found in the plankton, rather apart from the others. Chief among these features are the mode of formation of the coelom, which is by a rather modified form of enterocoely, and the fact that only during their development do they profess their coelomate and tripartite arrangement at all; subsequently the coelom is filled in with mesenchyme and a secondary cavity is formed, closely resembling that of the pseudocoelomate phyla such as the Nematoda. Chaetognaths, alone among the deuterostomes, are pelagic, and the adoption of such a habit has caused the similarities with the other deuterostomes to be almost entirely masked; so the phyletic position of the chaetognaths is far from certain. It has been suggested[134] that they lie at the base of the line leading towards the echinoderms and chordates.

The remaining three phyla, Pogonophora, Hemichordata and Echinodermata, are united by possession of a number of features, among which we may mention the formation of the coelom by enterocoely, the retention of the blastopore, when present, as the site of the future anus (deuterostomy), and, in those with indirect development, a dipleurula-type larva at some stage. Also there is similarity in the configuration of the coelomic pouches. In all three there is said to be a heart vesicle (pericardial sac in some pogonophores, dorsal sac in echinoderms). The adult pogonophores, with their complete absence of alimentary system, are to be regarded as highly specialised offshoots of the line leading to the echinoderms. There is, though, the phenomenon of uptake of soluble foodstuffs from the sea-water which both echinoderms and pogonophores show markedly (p. 44). Very little is known about the pogonophore larva, beyond the fact that it is bilateral, has no blastopore and shows some signs of an incipient gut which never develops very far. The larvae of the other two have been well studied, however, and show fairly conclusively that the hemichordates and echinoderms are closely linked. The hemichordate larva, called a *tornaria*, has the same basic structure as the dipleurula of an echinoderm, with the difference that there is an extra ciliated band posteriorly, the telotroch. In other details there is remarkable similarity, such as the presence of a hydroporic canal associated with the left side in both. This similarity is confirmed in adult features, too, and Grobben[133] has shown the principal transformations necessary to convert a pterobranch hemichordate into an echinoderm, a derivation which requires chiefly the reduction of the right middle coelom (hydrocoel) and the growth round the oesophagus of the remaining hydrocoel pouch; the hemichordate cephalic shield he considers as the stalk and attachment disk of crinozoans and asteroids respectively (Fig. 27).

There have been several attempts to construct a hypothetical

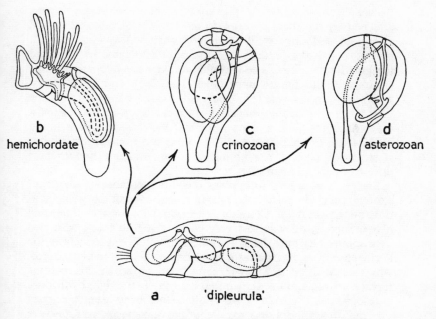

b
hemichordate

c
crinozoan

d
asterozoan

a   'dipleurula'

Fig. 27 Possible derivation of pterobranch hemichordates,
crinozoan and asterozoan echinoderms from the ancestral 'dipleurula'
(modified after Bather and Grobben)
Such a scheme is not advocated here.

a The *dipleurula*, drawn upside down when compared with Fig. 24*a*,
showing the paired axohydrocoels, each with its own hydropore,
and paired somatocoels.

b Pterobranch, showing migration of anus, single anterior coelom
and paired middle coelom, with branches to the tentacles.

c Crinozoan, showing changes envisaged by Bather in bringing the
mouth to the top. One axohydrocoel is lost; the axocoel part of
the other forms the parietal canal, lost later in ontogeny.

d Asterozoan, in which less drastic torsion has occurred.

common ancestor of all the echinoderms and the hemichordates,
the best known of which is that of Semon[139], developed later by
Bather[1]. Bather saw the ancestor as resembling the *dipleurula*
larval stage, but with a full complement of coelomic pouches on
either side of the gut, as in Fig. 27*a*; it was a creeping animal, with
anterior sense organ and ventral mouth and anus. Each axohydrocoel (p. 150) opened by a hydropore dorsally. A postulate of

Bather's theory was that the ancestor at some stage became attached by the front end (as do modern larval crinozoans and asteroids) and underwent torsion in such a way that the left axohydrocoel was reduced and the right one came to surround the gut. Bather shows the way this might have happened in the crinozoan, with mouth uppermost (Fig. 26c), and in asteroids, with mouth to one side (Fig. 26d). Although the movement of the internal organs of the crinozoan on this theory looks drastic, a very important piece of supporting evidence for the idea is found in the fossil cystoid *Aristocystites* (p. 90 and Fig. 13b) which has a structure closely resembling that of Bather's hypothetical primitive pelmatozoan after torsion.

Bather is at a loss to explain the disappearance of the right axohydrocoel. In fact, nobody has given a satisfactory explanation of the mystery. Certainly, the symmetry of the dipleurula if it existed at all was lost very early, since no fossil with two hydropores has ever been found. Gemmill[37] has reviewed the occurrence of anomalous asteroid larvae with double hydropores, which occur in up to thirty-three per cent of individuals in some larval cultures, and has shown that later in life the right hydropore is resorbed; this curious state of affairs has been used to suggest that the situation did once exist, that the odd specimens found are a sort of throwback; but such an idea is completely untenable.

The subsequent development of more advanced echinoderms, with the mouth first formed on one side of the larva (Fig. 27d), was probably by a less drastic torsion, with the consequent lack of twisting of the gut, retained in the asteroids and ophiuroids. The resulting picture is not unlike that of another hypothetical ancestor of the echinoderms (but not shared with the hemichordates), the *pentactula*, first suggested again by Semon[139], but developed later by Bury[131]. In this theory the original echinoderm already has five tentacles, the lumina of which stem from a circum-oral coelomic vessel. To me, it seems simplest to regard the pentactula as a possible later stage of the ancestral echinoderm, the larva of which was the dipleurula, and as representing an ancestor already well within the confines of our definition of an echinoderm. But there seems to be sufficient evidence to show that there is no need to regard the original echinoderm (p. 22) as pentamerous. One should describe it as a sessile animal with the mouth directed upwards and exterior near the mouth. The anus opens to one side of the theca. The dipleurula-concept may represent a possible *larval stage* common to the echinoderms and hemichordates, but the original echinoderm was probably not a creeping animal at all but a sessile one, probably already armoured with a test. It most likely rested on or in the silt at the bottom of the ancient sea with its arms permanently extended but its prehensile tube-feet capable of extension and retraction to help the feeding process.

The echinoderms and the origin of chordates

It is appropriate that the last section in this book should attempt to relate the echinoderms to the great phylum of the Chordata. It may seem absurd to try to relate the adults of the two groups, yet at least two serious attempts have been made. Gislén[147] first suggested that the curious slit-like structures on one side of the homalozoan *Cothurnocystis* (p. 114 and Fig. 19c and d) resemble the gill-slits of an immediately post-larval Amphioxus (a primitive chordate), and may be the first appearance of gill-slits in evolution. There is no proof that they were gills at all: it is not known whether they connected with the gut internally as do chordate gill-slits, though their position on the body makes the suggestion attractive. Jefferies[85, 148] has gone further: he sees the whole homalozoan body in chordate terms: the feeding arm becomes a swimming organ and the oral region near its base becomes the site of a chordate-like brain (see also p. 111).

The idea of relating adult echinoderms to adult chordates must be put aside as untenable, or, at most, unlikely. How, then, did the fish-like organisation of the primitive adult chordate arise from the invertebrate stock? How did the unique features of dorsal nerve cord, ventral heart, pharyngeal gill-slits and noto-chord, followed by vertebral column, jaws and so on, come into being? The clue to these questions probably lies in an interesting embryological phenomenon known as *neoteny*, in which, by the acceleration of development of the gonads, an animal becomes sexually mature while still retaining the larval body form. In other words, the animal never undergoes metamorphosis. Although this process can be induced phenotypically (that is, by an influence of the environment) it also seems highly probable that it has been utilised during evolution to bring about various new groups, some of which have shown tremendous potentiality for exploiting new habitats thus opened to them. It is unlikely that a new group could *suddenly* be produced in this way by one drastic genetic change;

but more probable that the onset of sexual maturity is *gradually* brought forward during development.

We have seen in Chapter 13 that the early dipleurula larva of echinoderms has a ciliated band running along its sides near the dorsal surface; the bands from the two sides join just in front of the anus and also near the mouth, where a continuation of each side, the *adoral* portion, passes a little way into the larval oesophagus on its ventral side. The point to be borne in mind here is that every ciliated band, if all its cilia are to work in concert for locomotion, must be underlain by some sort of nerve tract. So we have, in the auricularia, a strong aggregation of nerve tissue near the dorsal surface on each side. The theory of chordate origin first propounded by Garstang[145] in 1894 proposes that these two bands moved further towards the mid-line and rolled in on themselves to form a dorsal nerve tube. The reason for such a movement may have been that strong lateral blocks of swimming muscles were required for locomotion in place of the rather weak method of ciliary beat, particularly if selection was acting to increase the time the larva was in the plankton and probably also its final size. Such an increase in size would immediately raise two problems: first, there would be a need for extra support, particularly to maintain a constant length, and this may well have been solved by the incorporation of a rod, the *notochord*, down the back of the animal. We can be fairly certain that this rod consisted originally, as it does now, of vacuolated cells whose turgidity provided the necessary strength, rather as a plant is supported by the pressure of fluid in its cells. The second problem was that with increase in size and activity, gaseous exchange by diffusion over the whole body surface could no longer cope with the respiratory needs of the animal, so gill-slits were provided in the first part of the gut, the *pharynx*, so that water could be brought into the mouth, filtered to collect food particles and passed across a surface in close contact with some sort of vascular fluid. This current was probably helped into the gut by the retained part of the ciliated band inside the mouth, now called the *endostyle*. A larva of this type would be well suited to exploit the rich harvest of plankton near the sea-surface, so it would be to its advantage to delay metamorphosis into a sessile adult for as long as possible —for ever, if it could. And this is where the process of neoteny comes in: selection would favour the gradual advance in the onset of sexual maturity, so that finally an adult animal existed, the *protochordate*, which had a dorsal neural tube, gill-slits, a noto-chord and an endostyle. Further increase in size of this form would mean that the notochord became inadequate to counter the stresses of muscular movement, and a bony structure, the verte-bral column, was incorporated around it.

So we have now reached a stage recognisable as fish-like. There is no general agreement as to which living forms among the primitive chordates are closest to the ancestral pattern[141], but the views of Berrill[140], who regards the tadpole of an ascidian (sea-squirt) as occupying this position, appear to find widest acceptance. But whatever are the finer points concerning the course actually taken during the early evolution of the uniquely successful chordates, it is quite likely that the early echinoderms represented the springboard from an invertebrate condition to the vast potentialities of the vertebrates.

A CLASSIFICATION OF THE EC...

This list includes all the genera men...
† extinct groups or g...
(Br.) those occurring in B...

SUB-PHYLUM C...

CLASS CRINOIDEA
Order Inadunata† *Ramsey...*
Order Flexibilia† *Protaxo...*
Order Camerata† *Platyc...*
Order Articulata *Pentac...*
(Br.), *Notocrinus, Ison...*

CLASS CYSTOIDEA†
Order Diploporita *Sphaeronites,...*
locystis, Fungocystis
Order Rhombifera *Macrocystella, Cheiroc...*
phaerites, Pleurocystites, Echinosphaerites, Echin...
Cystoblastus, Staurocystis

CLASS EOCRINOIDEA† *Eocystites, Lichenoides*

CLASS PARACRINOIDEA† *Comarocystites*

CLASS PARABLASTOIDEA† *Blastoidocrinus*

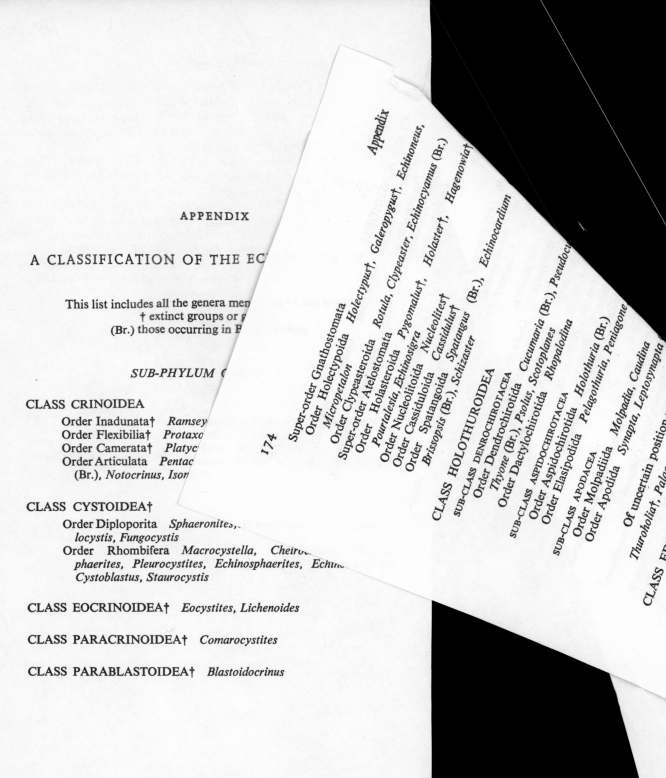

Super-order Gnathostomata
Order Holectypoida *Holectypus†, Galeropygus†, Echinoneus,*
Micropetalon
Order Clypeasteroida *Rotula, Clypeaster, Echinoneus,*
Super-order Atelostomata
Order Holasteroida *Pourtalesia, Echinosigra*
Pourtalesia, Echinosigra
Order Nucleolitoida *Nucleolites†*
Order Cassiduloida *Cassidulus†*
Order Spatangoida *Spatangus* (Br.), *Schizaster*
Brissopsis (Br.), *Schizaster* ... *Echinocardium*

CLASS HOLOTHUROIDEA
SUB-CLASS DENROCHIROTACEA
Order Dendrochirotida *Thyone* (Br.), *Psolus, Scotoplanes ... Cucumaria* (Br.), *Pseudo...*
Order Dactylochirotida *Rhopalodina*
SUB-CLASS ASPIDOCHIROTACEA
Order Aspidochirotida *Holothuria* (Br.)
Order Elasipodiida *Pelagothuria, Peniagone...*
SUB-CLASS APODACEA
Order Molpadiida *Molpadia, Caudina*
Order Apodida *Synapta, Leptosynapta...*

Of uncertain position within the...
Thuroholia†, Palaeocucumaria†

CLASS EDRIOASTEROIDEA†
Family Stromatocystitidae
Family Cyathocystitidae *C...*
Family Hemicystitidae
Family Agelacrinidae
Thresherodiscus
Family Edrioasteridae
Family Astrocystitid...

CLASS HELICOPLACOID...
CLASS OPHIOCISTIOID...
Volchovia
CLASS CYCLOCYST...

protochordate, which had a dorsal neural tube, gill-slits, a noto-chord and an endostyle. Further increase in size of this form would mean that the notochord became inadequate to counter the stresses of muscular movement, and a bony structure, the verte-bral column, was incorporated around it.

So we have now reached a stage recognisable as fish-like. There is no general agreement as to which living forms among the primitive chordates are closest to the ancestral pattern[141], but the views of Berrill[140], who regards the tadpole of an ascidian (sea-squirt) as occupying this position, appear to find widest acceptance. But whatever are the finer points concerning the course actually taken during the early evolution of the uniquely successful chordates, it is quite likely that the early echinoderms represented the springboard from an invertebrate condition to the vast potentialities of the vertebrates.

APPENDIX

A CLASSIFICATION OF THE ECHINODERMATA

This list includes all the genera mentioned in this book
† extinct groups or genera
(Br.) those occurring in British waters

SUB-PHYLUM CRINOZOA

CLASS CRINOIDEA

Order Inadunata†   *Ramseyocrinus, Hybocystis, Petalocrinus*
Order Flexibilia†   *Protaxocrinus, Ichthyocrinus*
Order Camerata†   *Platycrinites, Barrandeocrinus*
Order Articulata   *Pentacrinites†, Hyocrinus, Ptilocrinus, Antedon*
(Br.), *Notocrinus, Isometra*

CLASS CYSTOIDEA†

Order Diploporita   *Sphaeronites, Aristocystites, Eucystis, Dactylocystis, Fungocystis*
Order Rhombifera   *Macrocystella, Cheirocrinus, Glyptosphaerites, Pleurocystites, Echinosphaerites, Echinoencrinites Cystoblastus, Staurocystis*

CLASS EOCRINOIDEA†   *Eocystites, Lichenoides*

CLASS PARACRINOIDEA†   *Comarocystites*

CLASS PARABLASTOIDEA†   *Blastoidocrinus*

CLASS BLASTOIDEA†

Order Fissiculata *Astrocrinus, Codaster, Pterotoblastus, Thaumatoblastus*
Order Spiraculata *Nucleocrinus, Pentremites, Orbitremites, Eleutherocrinus*

SUB-PHYLUM ASTEROZOA

CLASS ASTEROIDEA

SUB-CLASS SOMASTEROIDEA *Chinianaster†, Villebrunaster†, Archegonaster†, Arthraster†, Platasterias*

SUB-CLASS EUASTEROIDEA
Order Platyasterida *Luidia* (Br.)
Order Phanerozonida *Petraster†, Hudsonaster†, Xenaster†, Astropecten* (Br.). *Podosphaeraster, Pectinaster*
Order Spinulosida *Porania* (Br.), *Solaster* (Br.), *Asterina* (Br.), *Anseropoda* (Br.), *Henricia* (Br.)
Order Euclasterida *Brisinga*
Order Forcipulata *Asterias* (Br.), *Heliaster, Leptasterias*

CLASS OPHIUROIDEA

Order Stenurida† *Pradesura, Eophiura, Palaeura, Aspidosoma, Stenaster, Taeniaster*
Order Oegophiurida *Lapworthura†, Ophiocanops*
Order Phrynophiurida *Ophiomyxa, Asteronyx, Asteroschema, Gorgonocephalus* (Br.)
Order Ophiurida *Ophiura* (Br.), *Ophiocomina* (Br.), *Ophiopsila* (Br.), *Ophiactis* (Br.), *Amphiura* (Br.), *Acrocnida* (Br.), *Amphipholis* (Br.), *Ophiotholia*

SUB-PHYLUM ECHINOZOA

CLASS ECHINOIDEA

SUB-CLASS PERISCHOECHINOIDEA
Order Bothriocidaroida† *Bothriocidaris*
Order Echinocystitoida† *Aulechinus*
Order Palaechinoida† *Palaechinus, Melonechinus, Lepidocentrus*
Order Cidaroida *Miocidaris†, Cidaris*

SUB-CLASS EUECHINOIDEA
Super-order Diademacea *Pygaster†, Echinothuria, Diadema*
Super-order Echinacea *Echinus* (Br.), *Psammechinus* (Br.), *Paracentrotus* (Br.), *Colobocentrotus, Toxopneustes*

Super-order Gnathostomata
  Order Holectypoida   *Holectypus*†, *Galeropygus*†, *Echinoneus*,
    Micropetalon
  Order Clypeasteroida   *Rotula, Clypeaster, Echinocyamus* (Br.)
Super-order Atelostomata
  Order Holasteroida   *Pygomalus*†,   *Holaster*†,   *Hagenowia*†,
    Pourtalesia, Echinosigra
  Order Nucleolitoida   *Nucleolites*†
  Order Cassiduloida   *Cassidulus*†
  Order Spatangoida   *Spatangus* (Br.), *Echinocardium* (Br.),
    Brissopsis (Br.), *Schizaster*

CLASS HOLOTHUROIDEA

SUB-CLASS DENROCHIROTACEA
  Order Dendrochirotida   *Cucumaria* (Br.), *Pseudocucumis* (Br.),
    Thyone (Br.), *Psolus, Scotoplanes*
  Order Dactylochirotida   *Rhopalodina*

SUB-CLASS ASPIDOCHIROTACEA
  Order Aspidochirotida   *Holothuria* (Br.)
  Order Elasipodida   *Pelagothuria, Peniagone, Psychropotes*

SUB-CLASS APODACEA
  Order Molpadiida   *Molpadia, Caudina*
  Order Apodida   *Synapta, Leptosynapta* (Br.), *Labidoplax* (Br.)

Of uncertain position within the holothuroids

Thuroholia†, *Palaeocucumaria*†

CLASS EDRIOASTEROIDEA†

Family Stromatocystitidae   *Stromatocystites*
Family Cyathocystidae   *Cyathocystis*
Family Hemicystitidae   *Hemicystites, Pyrgocystis*
Family Agelacrinidae   *Agelacrinus, Lepidodiscus,*
  Thresherodiscus
Family Edrioasteridae   *Edrioaster, Dinocystis*
Family Astrocystitidae   *Astrocystites*

CLASS HELICOPLACOIDEA†   *Helicoplacus, Waucobella*

CLASS OPHIOCISTIOIDEA†   *Sollasina, Euthemon, Eucladia,*
  Volchovia

CLASS CYCLOCYSTOIDEA†   *Cyclocystoides*

SUB-PHYLUM HOMALOZOA†

CLASS HOMOSTELEA   *Gyrocystis, Trochocystites*

CLASS HOMOIOSTELEA   *Dendrocystoides*

CLASS STYLOPHORA

Order Cornuta   *Ceratocystis, Cothurnocystis, Scotiaecystis*
Order Mitrata   *Mitrocystites*

Of uncertain position within the phylum

'Haplozoa'†   *Peridionites, Cymbionites*
Camptostromatoidea†   *Camptostroma*
Lepidocystoidea†   *Lepidocystis*

BIBLIOGRAPHICAL REFERENCES

The first section lists general works covering the whole phylum that have appeared since 1900; the second lists those works which deal with the special systems and features of echinoderms. Thereafter, the references are divided into sections corresponding to the chapters. All references are numbered consecutively, but each section except the first is arranged in alphabetical order of authors. The list has been kept to a minimum, and includes only those which the student requiring further information may find useful: it is not intended as a list of sources from which material has been taken.

GENERAL WORKS ON ECHINODERMS SINCE 1900,
IN CHRONOLOGICAL ORDER

1 Bather, F. A., Gregory, W. K. and Goodrich, E. S., 1900. 'The Echinoderma' in *A Treatise on Zoology*, Ed. E. Ray Lankester, London.

2 Ludwig, H. and Hamann, O., 1901–7. 'Echinodermen (Stachelhäuter)' in *Klassen und Ordnungen des Tierreichs*, Ed. Bronn, H. G., Berlin.

3 Delage, Y. and Herouard, E., 1903. 'Les Echinodermes' in *Traité de Zoologie concrete*, III, Paris.

4 MacBride, E. W., 1906. 'Echinodermata' in *The Cambridge Natural History*, Ed. Harmer and Shipley, London.

5 Koehler, R., 1924–7. *Les Echinodermes des Mers d'Europe*, Paris.

6 Mortensen, T., 1927. *Handbook of the Echinoderms of the British Isles*, Oxford University Press.

7 Cuénot, L., 1948. 'Echinodermes' in *Traité de Zoologie*, Ed., Grassé, P.-P., Paris.

8 Hyman, L. H., 1955. *The Invertebrates*, Vol. IV, Echinodermata, New York.

9 Clark, A. M., 1962. *Starfishes and their relations*, British Museum (Natural History), London.

10 Nichols, D., 1964. *Oceanogr. mar. Biol. Ann. Rev.*, *2*: 393–423. (Review of some recent research)

11 Boolootian, R. A. (Ed.), 1966. *Physiology of Echinodermata*, New York.

12 Moore, R. C. (Ed.), 1966–7. *Treatise on Invertebrate Paleontology*, Parts S and U, Echinodermata, Vols. 1 and 3, Kansas.

13 Millott, N., 1967. *Echinoderm Biology, Symp. zool. Soc. Lond.*, *20*, London.

SPECIAL FEATURES OF ECHINODERMS

14 Boolootian, R. A. and Giese, A. C., 1958. *Biol. Bull. mar. biol. Lab., Woods Hole, 115:* 53–63. (Coelomic corpuscles)

15 Heyde, H. C. van der, 1922. *On the Physiology of Digestion, Respiration and Excretion in Echinoderms.* Den Helder (Thesis)

16 Nichols, D., 1966. In *Physiology of Echinodermata*, Ed. Boolootian R. A., New York, Ch. 9. (Water-vascular system)

17 Nichols, D., 1969. 'Echinoderms' in *Practical Invertebrate Zoology*, Ed. Dales, R. P., London. (Practical and experimental guide)

18 Smith, J. E., 1965. 'Echinodermata' in *Structure and function in the nervous system of invertebrates*, Ed. Bullock, T. H. and Horridge, G. A., London. (Nervous system)

19 Ursin, E., 1960. *Meddr Kommn Danm. Fisk. -og Havunders.*, *2:* 1–204. (Quantitative study of North Sea echinoderms)

THE CRINOIDS

20 Chadwick, H. C., 1907. *L.M.B.C. Memoir, 15.* (*Antedon*)

21 Clark, A. H., 1915–50. *Bull. U.S. natn. Mus., 82* (Monograph of the existing crinoids)

22 Moore, A. R., 1924. *J. gen. Physiol., 6:* 281–8 (Nervous co-ordination in *Antedon*)

23 Moore, R. C., 1950. *Int. geol. Congr. Rep. 18th Sess. Gt. Brit., 1948, 12:* 27–53. (Evolution of crinoids)

24 Moore, R. C. (Ed.), in prepn. *Treatise on Invertebrate Paleontology*, Part T, Echinodermata Vol. 2, Kansas.

25 Regnéll, G., 1960. *Palaeontology, 2:* 161–79. (Lower Palaeozoic crinoids of Britain and Scandinavia)

THE ASTEROIDS AND OPHIUROIDS

26 Anderson, J. M., 1966. In *Physiology of Echinodermata*, Ed., Boolootian, R. A., Ch. 14. New York (Asteroid digestive system)

27 Binyon, J., 1961. *J. mar. biol. Ass. U.K., 41:* 161–74. (Salinity tolerance of *Asterias*)

28 Binyon, J., 1962. *J. mar. biol. Ass. U.K., 42:* 49–64. (Ionic regulation in *Asterias*)

29 Chadwick, H. C., 1923. *L.M.B.C. Memoir, 25.* (*Asterias*)

30 Davis, W. P., 1966. *Bull. mar. Sci. Gulf Caribb., 16:* 435–44. (Biology of phrynophiurid ophiuroids)

31 Fell, H. B., 1960. *Synoptic key to the genera of ophiuroids.* Victoria Univ., Wellington, N.Z.

32 Fell, H. B., 1963. *Phil. Trans. R. Soc.,* B, *246:* 381–435. (Phylogeny of starfishes)

33 Fell, H. B., 1966. *Oceanogr. mar. Biol. Ann. Rev., 4:* 233–45. (Echinoderm 'living fossils')

34 Ferguson, J. C., 1966. *Trans. Amer. microsc. Soc., 85:* 200–9. (Coelomocyte production by Tiedemann's bodies)

35 Ferguson, J. C., 1967. *Biol. Bull. mar. biol. Lab., Woods Hole, 132:* 161–73. (Uptake of exogenous amino acids by external tissues)

36 Fontaine, A. R., 1965. *J. mar. biol. Ass. U.K., 45:* 375–85. (Feeding of *Ophiocomina*)

37 Gemmill, J. F., 1914. *Phil. Trans. R. Soc.,* B, *205:* 213–94. (Development and adult structure of *Asterias*)

38 Gordon, I., 1929. *Phil. Trans. R. Soc.,* B, *217:* 289–334. (Skeletal development in *Leptasterias*)

39 Lavoie, M. F., 1956. *Biol. Bull. mar. biol. Lab., Woods Hole, 111:* 114–22. (How starfishes open bivalves)

40 Philip, G. M., 1965. *Nature, Lond., 208:* 766–8. (Ancestry of starfishes)

41 Smith, J. E., 1937. *Phil. Trans. R. Soc.,* B, *227:* 111–73. (Nervous system of *Marthasterias*)

42 Smith, J. E., 1950. *Phil. Trans. R. Soc.,* B, *234:* 521–58. (Nervous system of *Astropecten*)

43 Smith, J. E., 1950. *Symp. Soc. exp. Biol., 4:* 196–220. (Nervous mechanisms underlying behaviour in starfishes)

44 Spencer, W. K., 1951. *Phil. Trans. R. Soc.,* B, *235:* 87–129. (Somasteroids and early ophiuroids)

THE ECHINOIDS

45 Buchanan, J. B., 1966. *J. mar. biol. Ass. U.K., 46:* 97–114. (Biology of *Echinocardium*)

46 Bullock, T. H., 1965. *Amer. Zool.*, *5*: 545–62. (Conduction in the superficial nervous system of echinoids and asteroids)

47 Chadwick, H. C., 1900. *L.M.B.C. Memoir, 3*. (*Echinus*)

48 Cobb, J. L. S. and Laverack, M. S., 1966. *Proc. R. Soc.*, B, *164*: 624–66. (Anatomy and physiology of Aristotle's lantern)

49 Durham, J. W. and Melville, R. V., 1957. *J. Paleont.*, *31*: 242–72. (Classification of the echinoids)

50 Farmanfarmaian, A., 1968. *Comp. Biochem. Physiol.*, *24*: 855–63. (Haemal system, translocation and decontamination)

51 Harvey, E. B., 1956. *The American* Arbacia *and other Sea-Urchins*. Princeton. (General account of echinoid biology)

52 Hawkins, H. L., 1919. *Phil. Trans. R. Soc.*, B, *209*: 377–480. (Echinoid ambulacra)

53 Hawkins, H. L., 1943. *Q. Jl geol. Soc. Lond.*, *99*: lii–lxxv. (Evolution and habits of echinoids)

54 Kier, P. M., 1965. *J. Paleont.*, *39*: 436–65. (Evolutionary trends in Palaeozoic echinoids)

55 Kier, P. M. and Grant, R. F., 1965. *Smithson. misc. Collns.*, *149* (6). (Distribution and habits of reef-dwelling echinoids)

56 Lovén, S., 1833. *K. svenska VetenskAkad. Handl.*, *19*: 1–91. (Anatomy of *Pourtalesia* and other echinoids)

57 Millott, N. and Okumuru, H., 1968. *J. exp. Biol.*, *48*: 279–87. (Electrical activity in the echinoid radial nerve)

58 Millott, N. and Vevers, H. G., 1968. *Phil. Trans. R. Soc.*, B, *253*: 201–30. (Echinoid axial organ)

59 Mortensen, T., 1928–51. *A Monograph of the Echinoidea*, 5 vols. Copenhagen.

60 Nichols, D., 1959. *Phil. Trans. R. Soc.*, B, *242*: 347–437. (Morphology and habits of recent and fossil spatangoids)

61 Otter, G. W., 1932. *Biol. Rev.*, *7*: 89–107. (Rock-boring echinoids)

62 Sandeman, D. C., 1965. *J. exp. Biol.*, *43*: 247–56. (Electrical activity in the radial nerve and ampullae)

THE HOLOTHUROIDS

63 Chun, C., 1900. *Aus den Tiefen des Weltmeeres*, Berlin. (Anatomy of *Pelagothuria*

64 Clark, H. L., 1907. *Smithson. Contr. Knowl.*, *35*: 6–231. (Apoda and Molpadonia)

65 Endean, R., 1957. *Q. Jl microsc. Sci.*, *98*: 455–72. (Cuvierian organs of *Holothuria*)

66 Fish, J. D., 1697. *J. mar. biol. Ass. U.K., 47:* 129–44. (Biology of *Cucumaria*)

67 Frizzell, D. L. and Exline, H., 1955. *Bull. Miss. Sch. Mines Metal., Tech. Series, 89:* 1–204. (Monograph of fossil holothuroid spicules)

68 Pawson, D. L. and Fell, H. B., 1965. *Breviora, 214:* 1–7. (Classification of den'drochirotes)

69 Pople, W. and Ewer, D. W., 1954–5. *J. exp. Biol., 31:* 114–26 and *32:* 59–69. (Myoneural physiology)

70 Théel, H., 1882. *Rep. Challenger Soc., 4.* (Elasipoda)

EXTINCT CRINOZOANS

71 Bates, D. E., 1968. *Palaeontology, 11:* 406–9. (*Ramseyocrinus*, the oldest crinoid)

72 Kesling, R. V. and Paul, C. R. C., 1968. *Contr. Mus. Paleont. Univ. Mich., 22:* 1–32. (Interpretation of porocrinids)

73 Moore, R. C. (Ed.), 1967. *Treatise on Invertebrate Paleontology, Part S*, Echinodermata Vol. 1. Kansas. (See relevant chapters on these groups)

74 Paul, C. R. C., 1968. *Palaeontology, 11:* 697–730. (Dichopore structure in cystoids)

EXTINCT ECHINOZOANS

75 Bather, F. A., 1915. *Geol. Mag., VI, 2:* 211–15, 259–66. (Morphology and bionomics of edrioasteroids)

76 Durham, J. W., 1967. *J. Paleont., 41:* 97–102. (Interpretation of helicoplacoid ambulacral structure)

77 Durham, J. W. and Caster, K. E., 1963. *Science, 140:* 820–2. (Helicoplacoids a class)

78 Hecker, R. F., 1938. *C. R. (Doklady) Acad. Sci. U.R.S.S., 19:* 425–7. (The ophiocistioid *Volchovia*)

79 Moore, R. C. (Ed.), 1967. *Treatise on Invertebrate Paleontology, Part U*, Echinodermata Vol. 3. Kansas. (See relevant chapters on these groups)

80 Regnéll, G., 1948. *Norsk geol. Tidsskr., 27:* 14–58. (Ophiocistioids)

81 Sollas, I. B. J. and Sollas, W. J., 1912. *Phil. Trans. R. Soc.*, B, *202:* 212–32. (Ophiocistioids a class)

82 Webby, B. D., 1968. *Palaeontology, 11:* 513–25. (*Astrocystites*, an 'edrioblastoid' allied to the edrioasteroids)

HOMALOZOANS, 'HAPLOZOANS' AND LESSER EXTINCT
GROUPS

83 Bather, F. A., 1926. Introduction to Withers, T. H., *Catalogue of the Machaeridia*. British Museum (Natural History). London. (Machaerids interpreted as echinoderms)

84 Gislén, T., 1930. *Zool. Bidr. Uppsala, 12:* 199–304. (Affinities between homalozoans and chordates)

85 Jefferies, R. P. S., 1968. *Bull. Br. Mus. nat. Hist., 16:* 245–339. (Homalozoans interpreted as 'calcichordates')

86 Moore, R. C. (Ed.), 1966–7. *Treatise on Invertebrate Paleontology*, Parts S and U, Echinodermata Vols. 1 and 3. Kansas. (See relevant chapters on these groups)

87 Whitehouse, F. W., 1941. *Mem. Qd. Mus., 12:* 1–28. ('Haplozoans')

PENTAMERY AND THE ECHINODERM SKELETON

88 Breder, C. H., 1955. *Bull. Am. Mus. nat. Hist., 106:* 173–220. (Occurrence and attributes of pentagonal symmetry)

89 Gordon, I., 1926. *Phil. Trans. R. Soc.*, B, *214:* 259–312. (Development of the skeleton of *Psammechinus*)

90 Nichols, D., 1967. *Nature, Lond., 215*; 665–6. (Pentamerism and the calcite skeleton)

91 Nichols, D. and Currey, J. D., 1968. In *Cell Structure and its Interpretation*, Ed. McGee-Russell, S. M. and Ross, K. F. A., London. (Nature of echinoderm calcite)

92 Okazaki, K., 1960. *Embryologia, 5:* 283–320. (Skeleton formation in the echinoid larva)

SPINES AND PEDICELLARIAE

93 Campbell, A. C. and Laverack, M. S., 1968. *J. exp. mar. Biol. Ecol., 2:* 191–214. (Responses of echinoid pedicellariae)

94 Fujiwara, T., 1935. *Annot. Zool. Japon, 15,* 62–9. (Poisonous pedicellariae of *Toxopneustes*)

95 Jensen, M., 1966. *Ophelia, 3:* 209–20. (Responses of echinoid spines and pedicellariae)

96 Millott, N. and Takahashi, K., 1963. *Phil. Trans. R. Soc.*, B, *246:* 437–67. (Reaction of echinoid spines to light)

97 Perrier, M. E., 1870. *Ann. Sci. nat., 13:* 5–81. (Pedicellariae of asteroids and echinoids)

98 Romanes, G. and Ewart, J., 1881. *Phil. Trans. R. Soc.*, B, *172:* 829–85. (Pedicellariae of asteroids and echinoids)

99 Takahashi, K., 1967. *J. Fac. Sci. Tokyo Univ., 11:* 109–35. (Anatomy, physiology and mechanics of the echinoid spine)

100 Uexküll, J. von, 1899. *Z. Biol.*, *37*: 334–403. (Physiology of pedicellariae)

THE TUBE-FEET

101 Bargmann, W. and Behrens, B., 1963. *Z. Zellforsch.*, *59*: 746–70. (Fine structure of the asteroid ampulla)

102 Cobb, J. L. S., 1967. *Proc. R. Soc., B*, *168*: 91–9. (Innervation of the starfish ampulla)

103 Cobb, J. L. S. and Laverack, M. S., 1967. *Symp. zool. Soc. Lond.*, *20*: 25–51. (Neuromuscular systems in ampullae)

104 Fechter, H., 1965. *Z. vergl. Physiol.*, *51*: 227–57. (Function of the madreporite)

105 Kerkut, G. A., 1953. *J. exp. Biol.*, *30*: 575–83. (Forces exerted by tube-feet)

106 Kerkut, G. A., 1954. *Behaviour*, *6*: 206–32. (Co-ordination of tube-feet)

107 Kerkut, G. A., 1955. *Behaviour*, *8*: 112–29. (Retraction of tube-feet)

108 Nichols, D., 1959a. *Q. Jl microsc. Sci.*, *100*: 73–88. (Tube-feet of *Echinocardium*)

109 Nichols, D., 1959b. *Q. Jl microsc. Sci.*, *100*: 539–55. (Tube-feet of *Echinocyamus*)

110 Nichols, D., 1960. *Q. Jl microsc. Sci.*, *101*: 105–17. (Tube-feet of *Antedon*)

111 Nichols, D., 1961. *Q. Jl microsc. Sci.*, *102*: 157–80. (Tube-feet of *Cidaris* and *Echinus*)

112 Paine, V. L., 1926. *J. exp. Zool.*, *45*: 361–66. (Adhesion of tube-feet)

113 Paine, V. L., 1929. *Amer. Nat.*, *63*: 517–29. (Tube-feet as autonomous organs)

114 Réaumur, M. de, 1712. *Hist. Acad. roy. Sci.*, *4*: 115–45. (First account of tube-feet)

115 Smith, J. E., 1937. *J. mar. biol. Ass. U.K.*, *22*: 345–57. (Structure and function of various tube-feet)

116 Smith, J. E., 1946. *Phil. Trans. R. Soc., B*, *232*: 279–310. (Starfish tube-foot/ampulla system)

117 Smith, J. E., 1947. *Q. Jl microsc. Sci.*, *88*: 1–14. (Action of asteroid suckered tube-feet)

118 Woodley, J. D., 1967. *Symp. zool. Soc. Lond.*, *20*: 75–104. (Ophiuroid tube-foot system)

LARVAL FORMS AND DEVELOPMENT

119 Chadwick, H. C., 1914. *L.M.B.C. Memoir, 22.* (Echinoderm larvae)

120 Chaet, A. B., 1967. *Symp. zool. Soc. Lond., 20:* 13–24. (Gamete-releasing substances)

121 Chia, Fu-Shiang, 1966. *Biol. Bull. mar. biol. Lab., Woods Hole, 130:* 304–315. (Brooding behaviour in *Leptasterias*)

122 Dawydoff, C., 1948. In *Traité de Zoologie*, Ed. Grassé, P.-P., Paris. (Development and larval forms)

123 Nyholm, K.-G., 1951. *Zool. Bidr. Uppsala, 29:* 239–76. (Development of *Labidoplax*)

124 Rees, C. B., 1953. *J. mar. biol. Ass. U.K., 32:* 477–90. (Larvae of spatangoids)

125 Seelinger, O., 1892. *Zool. Jb. Anat., 6:* 161–444. (Development of *Antedon*)

126 Thorson, G., 1950. *Biol. Rev., 25:* 1–45 (Larval ecology)

For a prolonged, and in places heated, correspondence on the phyletic importance of echinoderm larval forms between F. A. Bather, W. Gemmill, E. W. MacBride and T. Mortensen from 1920 to 1923, see *Nature, Lond., 107:* 132–3; *108:* 459–60; *108:* 529–30; *108:* 530; *110:* 801–7; *111*, 47; *111:* 47–8; *111:* 322–3; *111:* 323–4; *111:* 397.

PHYLOGENY OF THE ECHINODERMS

127 Fell, H. B. and Pawson, D. L., 1966. In *Physiology of Echinodermata*, Ed. Boolootian, R. A., New York. 1–48.

128 Heddle, D., 1967. *Symp. zool. Soc. Lond., 20:* 125–41. (Asteroid arm movement and origin)

129 Moore, R. C. (Ed.), 1966–7. *Treatise on invertebrate Paleontology*, Parts S and U, Echinodermata Vols. 1 and 3, Kansas. (See sections on phylogeny throughout, but particularly Ubaghs, G., pp. S43–S52)

INVERTEBRATE RELATIONS OF ECHINODERMS

130 Bergmann, W., 1949. *J. mar. Res., 8:* 137–176. (Sterols as a guide to invertebrate phylogeny)

131 Bury, H., 1895. *Q. Jl microsc. Sci., 38:* 45–136. (Pentactula theory)

132 Clark, A. H., 1922. *Smithson. misc. Collns., 72; 11:* 1–20. (Echinoderms as aberrant arthropods; see also review by Bather, F. A., 1922, *Nature, Lond., 109:* 640–1)

133 Gröbben, K., 1923. *Sber. Akad. Wiss. Wien, 132:* 263–90. (Echinoderm-hemichordate relations)

134 Hyman, L. H., 1959. *The Invertebrates,* vol. V, Smaller Coelomate Groups, New York. (Deuterostome relations)

135 Kerkut, G. A., 1961. *Implications of evolution,* London. (Invertebrate relationships)

136 Marcus, E., 1958. *Quart. Rev. Biol., 33:* 24–58. (Evolution of the animal phyla)

137 Nichols, D., 1967. *Symp. zool. Soc. Lond., 20:* 209–29. (Sipunculoid origin of echinoderms)

138 Nichols, D., in press. In *The Invertebrate Panorama,* Ed. Harrison-Matthews, L. and Smith, J. E., London.

139 Semon, R., 1888. *Jena Z. naturw., 22:* 175–309. (Dipleurula and pentactula theories)

ECHINODERMS AND THE ORIGIN OF CHORDATES

140 Berrill, N. J., 1955. *The Origin of Vertebrates,* Oxford.

141 Bone, Q., 1960. *J. Linn. Soc. (Zool.), 44:* 252–69. (Origin of chordates from hemichordates)

142 Carter, G. S., 1957. *Syst. Zool., 6:* 187–92. (Not paedomorphosis)

143 Fell, H. B., 1940. *Nature, Lond., 145:* 906–7. (Origin of vertebrate coelom)

144 Fell, H. B., 1948. *Biol. Rev. 23:* 81–107. (Echinoderm embryology and the origin of chordates)

145 Garstang, W., 1894. *Zool. Anz., 17:* 122–5. (Echinoderms to chordates by paedomorphosis)

146 Garstang, W., 1928. *Q. Jl microsc. Sci., 72:* 51–187. (Position of tunicates in echinoderm-chordate line)

147 Gislén, T., 1930. *Zool. Bidr. Uppsala, 12:* 199–304. (Affinities between homalozoans and chordates)

148 Jefferiês, R. P. S., 1967. *Symp. zool. Soc., Lond., 20:* 163–208. (Homalozoan structure interpreted in chordate terms)

149 Moreland, B., Watts, D. C. and Virden, R., 1967. *Nature, Lond., 214:* 458–62. (Phosphagens in the echinoderm-chordate line)

INDEX

Page-numbers in **heavy type** refer to illustrations